Goethe's Botanical Writings

GOETHE ON THE ROMAN CAMPAGNA

From an oil painting by his friend Tischbein. During Goethe's sojourn in Italy, 1786–1788, he evolved his theory of plant metamorphosis.

Goethe's Botanical Writings

Translated by
BERTHA MUELLER

With an Introduction by
CHARLES J. ENGARD

OX BOW PRESS
Woodbridge, Connecticut

1989 reprint published by:
OX BOW PRESS
P.O. Box 4045
Woodbridge, Connecticut 06525

© Copyright, 1952, by the University Press of Hawaii
First published by the University Press of Hawaii, 1952.

The paper used in this publication meets the minimum requirements of American National Standard for Information Sciences—Permanence of Paper for Printed Library Materials, ANSI Z39.48-1984.
∞

Library of Congress Cataloging-in-Publication Data

Goethe, Johann Wolfgang von, 1749–1832.
  [Essays. English. Selections]
  Goethe's botanical writings / Johann Wolfgang Goethe ; translated by Bertha Mueller ; with an introduction by Charles J. Engard.
    p.   cm.
  Reprint. Originally published: University of Hawaii Press, 1952.
  Bibliography: p.
  1. Botany—Morphology. 2. Goethe, Johann Wolfgang von, 1749–1832—Knowledge—Botany. I. Title.
QK641.G63A5   1989
580—dc19   89–3120
                                                                                       CIP

ISBN 0–918024–69–2
ISBN 0–918024–68–4 (pbk.)

Printed in the United States of America

Dedicated to
CHARLES J. ENGARD

At present, by virtue of ever-widening experience and ever-deepening philosophy, much has become accessible that was not available to me and others when the following pages were written. Should they seem superfluous, let their content be looked upon from the historical viewpoint, as a chronicle of quiet, steadfast, and consistent effort.—GOETHE

Translator's Preface

THE INADEQUACIES of translations have been endlessly discussed: inadequacies stemming from disparities between historical periods and national cultures, between the rhythms, words, phrases, and figures of speech even of cognate languages, between the genius of original writers and the pedestrian qualities of translators—disparities leading to loss of that perfect fusion of style and content which is the mark of great writing. Of these and countless other inadequacies translators themselves are only too aware. Guilt-ridden after futile search for the ideal word, resigned instead to a choice of the lesser evil, they are not likely to follow the example of Luther, in his famous open letter, hurling defiance at the critics of his Bible translation. Lesser lights, though they may derive some comfort from his bitter taunts, are content with explaining the genesis of their specific translations, with setting forth special difficulties encountered and ways in which these were surmounted.

The present translation was undertaken at the suggestion of my colleague, the late Dr. Charles J. Engard, of the botany department. Shortly after completing translation of "The Metamorphosis of Plants," we learned that *Chronica Botanica* was soon to publish a translation of the same essay, by Professor Agnes Arber of Cambridge, England. However, on being informed that she was not contemplating further translation of Goethe's botanical works, and at her gracious urging, we decided to complete our project.

In contrast to difficulties involved in translating, let us say, from Chinese into English, the ones encountered in translating from German to English are obviously minor. They are, nevertheless, real. Chief among them is the greater compactness of German, necessitating substitution of whole phrases and clauses for compound nouns and split-adjective constructions. Such expansion, together with shifts in word order entailed by the transposed order of dependent clauses in German, will often erase all antecedents and demand complete reconstruction of the sentence. Yet it has been my objective throughout to achieve not merely idiomatic expression but faithful translation as well. In some of the essays, dictated by Goethe and insufficiently edited after transcription, vagueness of phrasing frequently made it difficult indeed to maintain the desired balance between unwarranted freedom and foolish awe.

The random order of the essays in Goethe's morphological journal has been dropped in favor of organization by content, the original order being indicated in the bibliographical notes. Within the organizational headings, an effort was made to present the essays chronologically by date of composition, although this plan could not be followed throughout, either because exact dates were unknown or because the chronological order did violence to the subject matter. Some essays without direct reference to botany, those comprising the section entitled "On General Theory," were included as being representative of the basic viewpoint of the botanical work.

The infrequent starred footnotes are Goethe's; the remaining ones, with the usual superior numerals, are my own. Some of them were suggested by Bölsche's commentaries, others by Dr. Engard, but for the most part they represent material readily available in encyclopedias and numerous Goethe biographies. A chronology of principal events in Goethe's life has been included, inasmuch as, by his own admission, he omitted "such superfluous things as dates" from the essays on the history of his plant studies. Plates and illustrations were adapted from Wilhelm Troll's *Goethes Morphologische Schriften* (Eugen Diederichs Verlag in Jena), with the permission of the publishers.

To Professor Maria Hörmann, friend and colleague of long standing, I should like to acknowledge my debt for willing help at all times with difficult passages and for graciousness in the unrewarding role of sounding board as the translator engaged in thinking aloud; to her husband, Dr. Arthur Hörmann, for various valuable suggestions; to Beatrice Krauss, of the Hawaiian Pineapple Research Institute, for ready assistance with botanical terms, especially after Dr. Engard's death; to several members of the University of Hawaii Press—to Thomas Nickerson, director, for gentle prodding and larger forms of encouragement; to Aldyth Morris for long and pleasant hours of collaboration during editing of the manuscript; to William Ellis for a book design worthy of a Goethe who was not insensitive to the physical appearance of books and who has recorded his pleasure on the publication of his essay on the metamorphosis of plants "elegantly printed in Roman letters." Finally, I should like to acknowledge my debt to Dr. Engard for suggesting a project holding such challenge and interest, and for infinite tact and patience in explaining technical details to a botanically unversed Germanist.

BERTHA MUELLER

University of Hawaii
June, 1952

Table of Contents

Translator's Preface vii
Introduction 3

ON MORPHOLOGY

Formation and Transformation 21
Metamorphosis of Plants 30
      Introduction 31
   I. Concerning the Seed Leaves 33
  II. Development of Stem Leaves from Node to Node . . 37
 III. Transition to Inflorescence 42
 IV. Formation of the Calyx 42
  V. Formation of the Corolla 44
 VI. Formation of Staminal Organs 47
VII. Nectaries 48
VIII. Additional Notes on the Staminal Organs 55
 IX. Formation of the Style 57
  X. The Fruits 61
 XI. The Proximate Hulls of the Seed 65
 XII. Recapitulation and Transition 67
XIII. Buds and Their Development 68
XIV. Formation of Compound Flowers and Fruits . . . 69
 XV. The Perfoliate Rose 73
XVI. The Perfoliate Pink 73
XVII. Linné's Theory of Anticipation 74
XVIII. Recapitulation 76
Metamorphosis of Plants—Second Essay 78
An Attempt to Evolve a General Comparative Theory . . . 81
Preliminary Notes for a Physiology of Plants 85
Later Studies and Collections 96
Pollination, Volatilization, and Exudation 105
Increasing Difficulty of Botanical Instruction 114
Remarkable Healing of a Badly Injured Tree 115
Problems 116
An Unjust Demand 118
Book Reviews 119

The Spiral Tendency 127
On the Spiral Tendency in Plants 131

ON HIS PLANT STUDIES

The Author Relates the History of His Botanical Studies 149
Genesis of the Essay on the Metamorphosis of Plants 165
History of the Manuscript 167
History of the Brochure in Print 170
My Discovery of a Worthy Forerunner 176
Three Favorable Reviews 181
Other Friendly Overtures 182
Notes for an Essay on Plant Culture in the Grand Duchy of Weimar 186
An Analogous Procedure 191
The Influence of My Publication 195

ON GENERAL THEORY

Propitious Encounter 215
Indecision and Surrender 219
The Objective and the Subjective Reconciled
   by Means of the Experiment 220
Experience and Science 227
Influence of the New Philosophy 228
Intuitive Judgment 232
The Creative Urge 233
Considerable Assistance from One Ingeniously Chosen Word . . 235
Analysis and Synthesis 238
Excursus 240
Friendly Gesture 240
Plea for Unity and Cooperation 241
Nature (A Fragment) 242
Commentary on "Nature" 244

Biographical Notes 249
Bibliographical Note 255
Selected Bibliography 258

Introduction

Higher natures have the gift of always seeing the finite world symbolically.—HEBBEL

INTRODUCTION

WHEN JOHANN WOLFGANG GOETHE arrived in Weimar in the darkness of predawn on November 7, 1775, he was already famous as a literary figure, but none could have foretold the prominent part he was also destined to play in the rise of a great scientific era.

With *Götz von Berlichingen* in 1773 he had established himself as the leading representative of the Storm and Stress movement then at its height in Germany; and with *Werther* in 1774, which created a sensation throughout Europe, he had established his reputation abroad.

Dressed in the costume of his own Werther, the young and handsome Goethe captivated the court circle of Weimar. His friend Karl Knebel wrote that Goethe "rose like a star in the heavens. Everybody worshipped him, especially the women." The Dowager Duchess Amalia, fascinated by his talent and exuberance, fell completely under his charm. Yet this was the man who later was to have a genus of plants *(Goethea)* and a mineral (goethite) named for him; who was to coin and be the first to use the word morphology; who was to contribute to our understanding of the physiology of color; who was to rediscover and describe the intermaxillary bone in man, propound the vertebral theory of the skull, formulate a concept in botanical morphology that persists to this day, discover the volcanic origin of a mountain, establish the first system of weather stations; who was to be among the first to use the comparative method in biology, to make the first systematic classification of minerals; and, finally, was to come unwittingly close to achieving the fundamental concept of organic evolution.

The year of Goethe's arrival in Weimar was a critical one politically throughout the western world. In America the battles of Concord and Lexington had been fought in April, and war had been declared on England in July. In France sentiment for the rights of man was growing; Louis XVI, destined to plunge France into bankruptcy, had become king the year before. The storming of the Bastille was fourteen years away.

In science the pneumochemists were active, and only in March of 1775 had Priestly communicated to the Royal Society in London his discovery of "dephlogisticated" air (oxygen). In the same year Black discovered "fixed" air (carbon dioxide). In America Benjamin Franklin had studied the Leyden jar, and had defined "positive" and "negative" electricity. Biologists of the time were concerned with the questions of Special Creation and of fixity versus mutability of species. There was as yet no comparative morphology—indeed, the word had not even been coined. In botany the great Linnaeus' works had completely eclipsed the anatomy of Grew and Malpighi of the previous century. Wolff's *Theoria Generationis* had been published in 1759, but this was unknown to a Goethe as yet not interested in plants. Koelreuter, some ten years before, had performed the first hybridizations with plants at Karlsruhe, but few people knew of his experiments.

Thirty years were to pass before Dalton would define the atom and many more before Mendel would found the science of plant breeding. The cell theory was unknown, and the concept of the gene lay far in the future. The greatest concept in biology, some say in the thinking of man, was yet to come: the theory of organic evolution and the descent of man. Alchemy was still being experimented with; even Goethe himself in a home laboratory had dabbled in alchemy. Now in 1775, however, he was developing a keen interest in biology.

Although Goethe had pursued the study of law, first at the university of Leipzig, later at the university of Strassburg where he obtained his licentiate, he had also included science among his studies. At Leipzig, where he lunched with students mostly in medicine and botany, he had become familiar with the names of such well-known scientists as Haller, Buffon, and Linnaeus. Later, at Strassburg, it was again mainly students of medicine with whom he fraternized at table. He was eager to learn more of this profession, and his dining companions supplied much of the incentive and interest. He attended lectures on anatomy, worked in the dissecting room, and heard clinical lectures on internal medicine and even on midwifery. Thus Goethe, though his main interest lay in literature, was no stranger to the methods of science.

In Weimar, captivated by the charm of the wooded countryside and especially by the splendid ducal gardens where he lived in the *Gartenhaus* given to him by the Duke of Saxe-Weimar, Goethe developed an active interest in plant life. This natural interest in plants was sharpened by his friendship with Dr. Wilhelm Heinrich Sebastian Buchholz, a local apothecary who kept a garden of medicinal herbs and other plants of special interest. Goethe, too, became interested in the project and together they built a veritable botanical garden. However, the ardor developing in Goethe for the new-found science of plants did not in any way detract from his concern with skeletal anatomy. He had been collaborating with Lavater, mystical poet and physiognomist, and had contributed the infor-

INTRODUCTION   5

mation on animal skulls in the latter's work on physiognomy. Later, in 1784, he rediscovered the intermaxillary bone in man[1]—of considerable importance at the time—and still later he enunciated the vertebral theory of the skull, a theory often mentioned in textbooks of comparative anatomy today.

The duties of his office as Privy Councillor and director of mines at Ilmenau took Goethe on frequent trips into the country, in the course of which he made friends with the many root gatherers who supplied apothecaries with materials for medicinal extracts and compounds. Thus Goethe was provided with a pleasant means of learning the natural habitats and the identities of the useful plants of the local flora.

Most influential in Goethe's early botanical education was his acquaintanceship, during the 1780's, with Gottlieb Dietrich, descendant of a Ziegenhain family that for generations had collected plants for medicinal purposes, had made herbaria, and were considered authorities on local flora generally. Goethe, who had befriended Gottlieb Dietrich, was greatly impressed with his knowledge of Linnaean botany. Their common interest in plants quickly developed into a close friendship, Goethe even taking Dietrich to the spa at Karlsbad where he went for a cure. At Karlsbad there was much collecting, identifying, and naming, and Goethe became a devotee of the Linnaean system.[2]

He carried with him on all of his field trips a single slender volume—his only botanical reference—which held between its covers Linnaeus' *Termini botanici, Fundamenta botanica,* and Gessner's *Dissertationes.* Also in his possession was a copy of Linnaeus' *Philosophia botanica,* which he says he studied daily.

Tormented by a restless desire to visit Italy, Goethe left Karlsbad incognito early one September morning in 1786 and proceeded across the Alps. As he left the Brenner Pass and journeyed down through northern Italy, his interest in nature was vigorously stimulated, he tells us, by the numerous larch and cembra nut trees, which drew his attention especially to climatic influences. Everywhere on his trip to the south of Italy he was impressed by the variety and luxuriance of plant forms; but the greatest impression was made on him by the foreign vegetation in the Botanical Garden in Padua, where a "high broad wall with the fiery red bells

---

[1] It is often stated that Goethe discovered this bone, but evidence indicates that it had been found earlier by the Frenchman Vicq d'Azyr. The work of each was unknown to the other. Goethe's account was more detailed; he gave the bone its place in human anatomy and is usually cited as its discoverer. It is sometimes called the "Goethe bone."

[2] For Goethe's own account of his relationship with Dr. Buchholz and Gottlieb Dietrich, see pp. 152–155.

of *Bignonia radicans* glowed enchantingly"[3] before him, and where he saw growing in the open many rare plants which in Germany grew only in hot houses. Of all the plants in the garden, the one which influenced him most was the fan palm, for, he noted, its leaves exhibited a complete series of transition forms from the simple lance-shaped first leaves to the most complex fan type.[4] This observation seems to have been the nucleus from which the doctrine of metamorphosis was later to grow. Nature, in the fan palm, had shown clear, living proof that all of these lateral outgrowths of the plant were simply variations of a single structure— the leaf. Goethe says in retrospect, "Because they may be grouped under one concept, it gradually became clearer and clearer to me that the concept could also be valid in a higher sense: a challenge which hovered in my mind at that time in the sensuous form of a supersensuous plant archetype. I traced the variations of all forms as I came upon them. And thus, in Sicily, the final goal of my journey, the conception of the original identity of all plant parts had become completely clear to me; and everywhere I attempted to pursue this identity and to catch sight of it again."[5] He caught sight of it again in Palermo, where, while working on the plot of his play *Nausicaa*, he would stroll in the beautiful public garden, amid groves of lemon and orange trees, and myriads of plants he had never seen before. Again the variety of forms, and yet their similarity, called to mind the *Urpflanze*. "Seeing such new and burgeoning growth," he wrote at this time, "I returned to my old idea and wondered whether I might not chance upon my archetypal plant. . . . I tried to find wherein these many and diverse forms were different, and always I found them more like than unlike."[6] The scientist in Goethe had emerged victorious over the poet; the garden of King Alcinous yielded to the garden of Palermo and the metamorphosis of plants![7]

Goethe continued to develop his ideas, make observations, and assemble notes and sketches. In 1790, two years after his return from Italy, these ideas were published in book form by Ettinger of Gotha after

---

[3] See p. 161.
[4] The "Goethe palm" still stands in the Botanical Garden at Padua. See Agnes Arber, "Goethe's Botany," *Chronica Botanica*, X (Summer, 1946), 77. For Goethe's description of this palm, see p. 161 of this translation.
[5] See p. 162.
[6] Johann Wolfgang Goethe, *Gedenkausgabe der Werke, Briefe, und Gespräche* (Zürich: Artemis-Verlag, 1949), XI, "Die Italianische Reise," 291.
[7] *Nausicaa*, a play based on a part of Homer's *Odyssey*, was not completed beyond the early form written in Palermo. Incidents such as this have caused some biographers of Goethe to look upon his scientific activities as impediments to the production of his poetry.

INTRODUCTION 7

Goethe's regular publisher had refused the manuscript.[8] The title of this first publication was *An Attempt to Explain the Metamorphosis of Plants;* later publications carried the less humble title *The Metamorphosis of Plants*. The book with its theory of plant morphology was ignored by botanists and the public alike, for, as Goethe says: "the public ... demands that every man remain in his own field."[9] It was eighteen years before references to the metamorphosis of plants began to appear in botanical texts and other writings, and thirty years before it was fully accepted by botanists. It was not the first nor the last theory of plant morphology; it was, however, the most influential, and still dominates all other theories today.

An analysis of Goethe's concept of the metamorphosis of plants, and its parental philosophy of unity in nature, may well begin with this quotation: "In organic being, first the form as a whole strikes us, then its parts and their shape and combination."[10] This is a self-evident truth. First we recognize, then we analyze. At first glance we recognize an organism as a plant; upon closer inspection we see that the plant is comprised of many parts, somewhat alike, somewhat different. The doctrine of metamorphosis—the modern concept of homology—tells us that all lateral outgrowths are referable to one fundamental ideal structure, the leaf. Among Goethe's notes is the aphorism: "Everything is leaf, and through this simplicity the greatest diversity becomes possible."[11] Let us see, in brief, just what this means.

Beginning with the seed, let us visualize the development of any common plant with flowers. The first "leaflike" structures to appear on the germinating seedling are the cotyledons—the word, derived from the Greek and meaning a hollow or cavity, has no reference to leaf—which may be fleshy and quite unleaflike as those of the bean, or thin and leaflike as those of the castor oil plant and many garden vegetables. One can see the veins and readily agree that they are "modified" leaves.[12] The formations that follow on the stem are commonly known as "foliage" leaves, and are thought of, perhaps quite erroneously, as "true" leaves. As Goethe has noted, production of this kind of leaf may be maintained

---

[8] For Goethe's account of this episode, see pp. 168–169.
[9] See p. 169.   [10] See p. 86
[11] Goethe used the word *Blatt*, which applies to the foliage leaf; but Goethe's *Blatt* is not any particular leaf as commonly seen, but a leaf as a type or idea, and should perhaps have been designated by the word *Urblatt*, or leaf prototype.
[12] Because one "conceives" the form of a leaf as the common bifacially expanded, petiolate, green-colored organ, it is the reference organ in this discussion. This is the foliage leaf.

indefinitely, under certain conditions of nutrition, and the formation of flowers prevented.[13]

In certain plants the growing tip of the stem is protected by a group of small leaflike structures known as bud scales or cataphylls. It is not difficult to find cataphylls, such as those of the horse chestnut, which show clearly all degrees of gradation from foliage leaf to the most reduced bud scale. Here the similarity to the leaf is easily recognized.

In the formation of an ordinary perfect flower—one with all its parts—there is, in ascending order, first a group of green, leaflike sepals. Next are the petals, commonly gaudy in color, and usually lacking in green pigment, chlorophyll, which is characteristic of foliage leaves. Next are the few to many stamens, whose terminal structures, the anthers, bear pollen which finally produces the sperm nuclei. Last in this series of modifications are the pistils, the most completely transformed of the fundamental leaf. These comprise the fruit, which may be fleshy or dry, single or fused into groups. A good example of this type of metamorphosed leaf is the pod of the pea or bean plant. When the pod is opened longitudinally and flattened out, there is before us, according to Goethe's theory, a leaf bearing the structures of reproduction—the seeds—along its margins. The mature pod, therefore, is a leaf with its two edges folded together. The composite cases, Goethe tells us, ". . . consist of several leaves grouped around a central point, their inner parts facing each other and their margins uniting."[14] The pistils, thus constituted of one or more sections, or carpels as they are now called, represent to Goethe the most perfect type of leaf.[15]

These homologies afforded to Goethe three possibilities of metamorphosis: (1) *regular,* or progressive; (2) *irregular,* or retrogressive; and (3) *accidental.* The last are malformations brought about by insects or other agents, and are not further discussed. Regular metamorphosis proceeds in the manner described above; irregular metamorphosis is a "backward" one from the point of view of the normal, as, for example, when stamens assume the structure of petals. It was clear to Goethe, however, that as far as metamorphosis was concerned it made no difference whether a petal is a transformed leaf, or a leaf is a transformed petal, because "forwards and backwards, the plant is ever only leaf." Goethe visualized regular metamorphosis as proceeding through a series of alternate expan-

---

[13] See p. 42.
[14] See p. 64.
[15] Indication that the Goethean doctrine of metamorphosis is still widely held among botanists is the almost universal application to the stamen and pistil of the technical term *sporophyll,* which means simply "spore-bearing leaf."

INTRODUCTION

sions and contractions of organs, moving progressively toward more perfect types. The increased perfection is the result of a progressive "refinement" of the plant juices as they ascend the stem; the cruder materials are filtered out, so to speak, by the foliage leaves, then the bracts, and so on.[16]

The Metamorphosis of Plants was written in an unusual style, much different from that of even contemporary scientific writings. In some ways this was unfortunate, for it contributed to misunderstandings between Goethe and his contemporaries in science. His choice of style was understandable, however, for it was modeled after that of Linnaeus' *Fundamenta botanica,* a book of 365 aphorisms, with which Goethe was most familiar. Furthermore, many of the ideas suggested in the *Metamorphosis* were left undeveloped; much of it was written in a loose, cryptic manner, one that permitted interpretation and invited people to read into it more than was intended. Based partly on imagination, perhaps its greatest value is that it stimulates imagination. The theory it presents has been of great value in interpreting the morphology of plants, the science which he named.[17]

Although Goethe's was the first elaborated account of the morphology of plants, the first published attempt at a classification of the external parts of plants was probably that of Theophrastus in 4 B.C., who stated: "Now the primary and most important parts, which are also common to most, are these—root, stem, branch, twig."[18] It is probably natural that this earliest analysis should make these plant parts, in Theophrastus' words, members corresponding to the members of animals. Indeed, he stated also that "many plants shed their parts every year even as stags shed their horns, birds . . . their feathers, four-footed beasts their hair; so that it is not strange that parts of plants should not be permanent, especially as what thus occurs in animals and the shedding of leaves in plants are analogous processes."[19]

The concept of the leaf as an important and fundamental part of

---

[16] See p. 40.

[17] Goethe coined and first used the word morphology, and is given credit for its publication. Günther Schmid, in the introduction to R. Magnus, *Goethe as a Scientist,* tr. Heinz Norden (New York: Henry Schumann, 1949), p. xiii, states that the literary priority is not Goethe's, however, but belongs to Carl Friedrich Burdach, an anatomist who introduced the term in 1800. Goethe, however, had used it in conversation and correspondence with Schiller several years earlier, and it was his authority that gave it popular and scientific currency.

[18] Theophrastus, *Enquiry into Plants,* tr. Arthur Hort (London: Heineman, 1916), p. 11.

[19] *Ibid.,* p. 7. For Goethe's opinions concerning attempts to draw analogies between plant and animal parts, see pp. 78–79.

the plant did not have its inception until 1759, when Caspar Friedrich Wolff produced his doctoral thesis, *Theoria Generationis,* mainly about animals. His thesis in regard to the morphology of the leaf was that the vegetative or foliage leaves become bud scales or flower parts through modifications brought about by a gradual decrease in "vegetative power." Wolff developed his theory as a result of microscopic examination of the growing points of stems. Goethe, not aware of this work until many years later, gave full credit to Wolff as a "worthy forerunner."[20]

Goethe's foliar interpretation of the carpel was put on a firmer, more scientific basis in 1827 by A. P. de Candolle, who emphasized the integrity of the "fundamental organs," stem, root, and leaf. De Candolle developed a "law of symmetry" which was quite compatible with Goethe's metamorphosis. This law dealt with the symmetrical disposition of leaves about the stem, a symmetry sometimes disguised by union of parts and inequalities of growth.

Since Goethe's time, plant morphologists have been concerned essentially with two lines of thought: (1) the interpretation of the leaf, and (2) the question of the homology of the floral parts with the leaf. Casimir de Candolle, the grandson of A. P. de Candolle, in 1868 advanced the idea that the leaf is a modified branch in which, in the ordinary bifacial leaf, the tip has atrophied at the apex and on the ventral side.[21] It is an interesting fact that Goethe made a similar suggestion about fifty years earlier, when he wrote: "When leaves divide, or rather when they advance from their original state to diversity, they are striving toward greater perfection, in the sense that each leaf has the intention of becoming a branch, and each branch a tree."[22] These thoughts, however, were developed no further.

Pursuing this idea in more thorough detail, Agnes Arber has developed the concept of the leaf as a partial-shoot with whole-shoot tendencies.[23] Here the *shoot* is conceived as a morphological unit, and the *stem* and *leaf* are discarded as discrete units: "The leaf is a partial-shoot, produced laterally from a parent whole-shoot . . . but never actually attaining this goal (of whole-shoot), since radial symmetry and the powers of apical growth and of self-reproduction are curtailed or inhibited." It is, of course, not pertinent here to go into the details of this enticing concept;

---

[20] See p. 176.
[21] Referred to in Agnes Arber, "The interpretation of leaf and root in the angiosperms," *Biological Review,* XVI (1941), 81–105.
[22] See p. 101.
[23] Arber, *op. cit.*

INTRODUCTION   11

it is sufficient to record the path into which morphological thought has proceeded.

Wilhelm Troll in Germany, in contradistinction to Arber, has returned to the strict Goethean type concept, recognizing the leaf, stem, and root as final, given units not subject to further analysis.[24] Concerning the question of the homology of the floral parts with the leaves, Troll has developed the concept strictly from Goethe's *Urpflanze* as the methodological center of morphological investigation, an approach which Arber has referred to as "Gestalt" morphology.[25] Later, Troll presented a foliar interpretation of the carpel, based on a theoretical modification (Goethe would say metamorphosis) of the peltate leaf.[26] He recognized three categories: (1) carpels with apparent peltation; (2) carpels with latent peltation; and (3) epeltate carpels. This concept is in opposition to that of Gregoire, who, on histological grounds, maintained that the floral parts are not referable to a type leaf but are *organa sui generis*.[27] Still a third concept is that of Thomas who suggested that the origin and evolution of the angiosperm carpel is from the Caytoniales.[28]

It is apparent that the Goethean doctrine that "all is leaf" is still very prominent in morphological thinking today, and that, although attempts are being made to redefine the approach to the question of homology, and the interpretation of the leaf is undergoing revision, the leaf concept probably will be the foundation upon which any new edifice will be constructed. At any rate, it can hardly be held with Sherrington that the doctrine of metamorphosis was, in the light of facts obtained in the progress of botany, "found not to be borne out," and consequently fell into "the doleful category of unlucky guesses."[29] It was neither doleful, nor unlucky, nor a guess.

The clue to the origin of the type concept, and ultimately the theory of metamorphosis, may be found in a few sentences written by Goethe in Padua, September 27, 1786, during the Italian journey: "Here in the new [plant] diversity that I am seeing for the first time the thought

---

[24] W. Troll, *Vergleichende Morphologie der höheren Pflanzen* (Berlin, 1937). Not seen by the writer, but referred to in Arber, *op. cit.*

[25] Agnes Arber, "The interpretation of the flower: a study of some aspects of morphological thought," *Biological Review*, XII (1937), 157–184.

[26] W. Troll, "Morphologie der schildförmigen Blätter," *Planta*, XVII (1932), 153–314.

[27] V. Gregoire, "La morphogenese et l'autonomie morphologique de l'apparcil floral," *La Cellule*, XXXXVIII (March, 1938), 1. Le Carpelle, 285–452.

[28] H. H. Thomas, "Palaeobotany and the origin of the angiosperms," *Botanical Review*, II (August, 1936), 397–418, and papers cited therein.

[29] Sir Charles Sherrington, *Goethe on Nature and on Science* (2d ed.; Cambridge: Cambridge University Press, 1949), p. 22.

becomes ever stronger that it might be possible to derive all plant forms from one. Only in this way could we truly determine genera and species, which until now has been done, in my opinion, in a very arbitrary manner. At this point I have reached an impasse in my botanical philosophy and I do not yet see how I will find my way out."[30] What was this impasse in his botanical philosophy?

The answer is to be found in Goethe's world philosophy. He was, by nature, an intuitive thinker, a type often referred to as a "picture thinker." For example, there is Goethe's much-quoted account of his conversation with Schiller: "We had reached his house; the conversation lured me in. I gave a spirited explanation of my theory of the metamorphosis of plants with graphic pen sketches of a symbolic plant. He listened and looked with great interest, with unerring comprehension, but when I had ended he shook his head, saying 'That is not an experience, that is an idea.' I was taken aback and somewhat irritated, for the disparity in our viewpoints was here sharply delineated.... Controlling myself, I replied, 'How splendid that I have ideas without knowing it, and can see them before my very eyes.' "[31]

The kernel of Goethe's philosophy lies in this metaphysical concept of nature : "The godhead is at work in the living, not in the dead; it is present in everything in the process of development and transformation, not in what has already taken shape and rigidified. Thus, reason in its strivings toward the divine, is concerned with growth and life, whereas understanding is concerned with putting to use what has already developed and grown torpid."[32] This thought of Goethe's may well be an adaptation of Kant's distinction between Understanding *(Verstand)* and Reason *(Vernunft)*.[33] Understanding is based on premises and hypotheses, and is Thought according to schematized categories, of value only in relation to experience; Reason (in the German sense), on the other hand, is distinct from Understanding in that it apprehends in one immediate act the entire system, and is Thought working without reference to application of concepts, and is therefore supersensuous. The latter, of course, leads to the archetype, the *Urphenomen.* This idea of the "becoming" of living nature—of even the rocks—is ever present in Goethe's writings. The reader will encounter frequently in the pages of this trans-

---

[30] "Italienische Reise," *op. cit.,* p. 65.
[31] See p. 217.
[32] Oswald Spengler writes that this sentence, part of a conversation with Eckermann, Feb. 13, 1829, comprises his own entire philosophy and that he would not change one word of it. *Der Untergang des Abendlandes* (München: Oskar Beck, 1923), I, 67.
[33] For Goethe's own reference to the Kantian philosophy, see pp. 228–234.

lation statements such as the following: "Form is something mobile, something becoming, something passing." "The doctrine of formation is the doctrine of transformation." "Metamorphosis is the key to the whole alphabet of nature." Here, then, we have insight into Goethe's philosophical outlook upon nature, which soon led him to the type-phenomena *(Urphenomene)* concept. Inevitably he must discard the Linnaean approach, for it was strictly analytical, fixed, and never-changing; it emphasized the constancy rather than the changeability of species. Goethe's approach, on the other hand, was synthetic, beginning with an idea and extending to analysis of particulars. "An idea," Goethe tells us, "is independent of space and time; experience is restricted within them. The simultaneous and the successive are therefore intimately bound together in an idea, whereas they are always separated in experience."[34] This was the impasse to which Goethe alluded. Linnaeus' system was inflexible, fixed in space and time; that of Goethe was without coordinates. Separating and counting, he tells us, was not in his nature.

It is evident that Goethe's philosophy was one which sought an approach to unity in nature, the whole from its parts, a one-world of nature. According to points of view in this respect, Goethe places investigators of natural science in one of four categories:[35] the utilizers, the fact-finders, the contemplators, and the comprehenders. In this order, their mental approaches to problems progress from "the seekers of things practical" through increasing inquisitiveness to the creative thinkers, the highest type, who, by proceeding from ideas, "simultaneously express the unity of the whole."

The great von Helmholz has said that the investigator must possess some of the intuition of the poet. Goethe, an artist and poet, proceeded from the synthetic to the analytic; from the supersensible idea, infinite and dimensionless, to the sensible, finite, easily counted, separated, and compared. This, the philosophic method of the comprehender, is summed up in these lines from Goethe's pen:

> If you would draw benefit from the whole,
> You must search for the whole in the smallest part.[36]

Goethe sought the *Urpflanze* and *Urtier,* but he was not what is sometimes called by scientists an armchair philosopher. The pages of this translation will reveal that Goethe was an assiduous observer, studying many

---

[34] See p. 219.
[35] See p. 92.
[36] From "Sprüche in Reimen," in Goethe's *Sämtliche Werke* (Leipzig: Max Hesses Verlag, n.d.).

kinds of natural phenomena in the minutest detail permitted by the methods of the day. Many sketches were made by his own hand, and many excellently detailed and accurate drawings, some in color, were executed under Goethe's direct supervision.[37]

From Naples Goethe wrote to Herder: "Furthermore, I must confide in you that I have approached closely to the secret of plant reproduction and organization, and that it is the simplest thing that one could possibly imagine. Under this [Italian] sky one can make the most wonderful observations. The chief point, where the germ is lodged, I have discerned quite clearly and beyond doubt. The rest I can also already see as a whole, only a few points still remaining to emerge more distinctly. The archetype plant as I see it will be the most wonderful creation in the whole world, and nature herself will envy me for it. With this model, and the key to it, one will be able to invent plants without limit to conform, that is to say, plants which even if they do not actually exist nevertheless might exist and which are not merely picturesque or poetic visions and illusions, but have inner truth and logic. The same law will permit itself to be applied to everything that is living."[38] Here we have the unity of nature in essence. The archetype plant, the *Urpflanze,* was now quite clear in Goethe's mind. It had occurred to him previously that "if all plants were not molded on one pattern, how could I recognize that they are plants?"[39] This is a thought that modern geneticists are beginning to ponder more, although until now they have been concerned almost entirely with specific and varietal differences among organisms. The genetic bases for variations and mutations are generally well understood, and the literature on the subject is voluminous, but the question why or even how organisms are similar—and they are usually easily recognized as similar—is a different matter. Goethe recognized that structures are similar in fundamental plan; he does not explain why, but in his philosophy of nature there were certain things to be accepted as given—the *Urphenomene.* To attempt to explain these "given" phenomena, such as the *Urpflanze,* is a hopeless task. We accept the universe as far as we know it, but we do not attempt to explain *why* it exists. It is difficult enough to understand *how!*

To complete our discussion of Goethe's botanical and morphological theories we must consider, at least briefly, the subject of Goethe and evolution. Much has been written concerning Goethe as a forerunner of

---

[37] The complete graphic material left by Goethe is being published under the auspices of the German Academy of Science (Leopoldina).
[38] To Herder, May 17, 1787. "Italienische Reise," *op. cit.,* p. 353.
[39] Written in Palermo, April 17, 1787. "Italienische Reise," *op. cit.,* p. 291.

Darwin; opinions range from aye to nay, and include all stages in between. A few authors, perhaps wisely, avoid the subject. Goethe did approach the idea of the theory of descent, especially in his later years, but it has remained a matter of opinion whether he was close to achieving the concept. J. Arthur Tompson has said that Goethe was the greatest poet of evolution; and Ernst Haeckel, in the generation following Goethe, was undoubtedly his foremost champion as a pre-Darwinian; in fact, he subtitled his famous *History of Creation,* published in 1876, as "a popular exposition of the doctrine of evolution in general, and of that of Darwin, Goethe, and Lamarck in particular." He states, in the text: "Jean Lamarck and Wolfgang Goethe stand at the head of all the great philosophers of nature who first established a theory of organic development, and who are the illustrious fellow workers of Darwin." Haeckel, by reading a great deal into Goethe's writings, makes a strong case for Goethe as a pre-Darwinian, basing his claim primarily on the *Metamorphosis of Plants* and on Goethe's belief that the form of an organism is the result of a balance between two forces: the centripetal, coming from within (Haeckel would read *inheritance*), which is responsible for unity of type, and the centrifugal, coming from without, which is an adaptive formative tendency, a tendency toward variation, brought about by the external environment.[40] Haeckel concludes that the metamorphosis of plants "consists not merely . . . in the changes of form which the organic individual experiences during its individual development, but, in a wider sense, in the transformation of organic forms in general. His idea of metamorphosis is almost synonymous with the theory of development."[41]

In contrast, there is the conclusion of Arber, far removed from the initial startling impact of the new evolution theory: "Goethe's theory was, however, in no sense intended to serve as an account of the past. . . . on purely morphological lines, and with the aid of such facts as were generally accessible in his day, he brought forward reasons for the probable truth of his theory. . . . Those who visualize the flower and its parts as having been derived phylogenetically from a vegetative shoot with foliage leaves, and who suppose that the carpel has at some stage gone through an actual process of infolding, are replacing a morphological idea, valid within its own sphere, by an *historical* picture, which belongs to a different world of thought, and for the validity of which the morphological concept provides no evidence. Morphological and phylogenetic concepts belong to different categories, and only confusion can come of the attempt

---

[40] See pp. 83–84.
[41] Ernst Haeckel, *The History of Creation,* tr., rev. E. R. Lankester (New York: Appleton, 1876). I, 90.

to reduce these categories to one."[42] Perhaps this last sentence states the crux of the problem. Kalischer, in Bielschowsky's biography of Goethe,[43] follows the Haeckel sentiment. Magnus appears to travel the middle road when he concludes that "Goethe never developed the evolutionary thought *to the point of an all-embracing scientific principle*[44] explaining the structure of the entire animal kingdom. The question of whether he must be regarded as a forerunner of Darwin therefore remains moot. . . . Goethe's own views of animal form were distinct and complete. Only in later years did Darwinian thoughts appear in them, and then only as secondary elements."[45] The italicized words imply that Magnus assumed that Goethe had at least an incomplete concept of evolution. It is true, as Magnus says, that every predecessor must be looked upon as a forerunner of those who follow. To a certain extent it is also true, in the words of Lewes, that "discoveries are, properly speaking, made by the Age, and not by men."[46] But it is also true that for every age there are men in the one preceding who lay at least part of the foundation. Goethe had the essence of the idea,[47] certainly in his later years, but he had no concise scheme or mechanism for it. Goethe's thinking was not mechanistic, and this, coupled with a lack of experimental details, precludes the possibility that Goethe in his generation could have presented a mechanism. The conclusion whether or not Goethe stands as a forerunner of Darwin seems

---

[42] Arber, "The interpretation of the flower," p. 175.

[43] "The Naturalist," by S. Kalischer, in Albert Bielschowsky's *The Life of Goethe,* tr. Wm. A. Cooper (London and New York: G. P. Putnam's Sons, 1908), III, 81–134.

[44] Italics mine.

[45] Magnus, *op. cit.,* pp. 117 f.

[46] George Henry Lewes, *The Life of Goethe* (Everyman's Library ed.; London: Dent and Sons, 1908), p. 368.

[47] See the essay, "An Attempt to Evolve a General Comparative Theory," pp. 81–84. Even more direct statements are to be found in some of his other writings, as for instance, this (p. 25): "If one looks at plants and animals in their most rudimentary stages, they are scarcely distinguishable from one another. Such nuclear organisms—whether rigid, mobile, or semi-mobile—are just barely perceptible to our senses. Whether these first beginnings could be conclusively traced in opposing directions, to the plant through light and to the animal through darkness, I do not make bold to decide. . . . This much we can say: creatures, emerging gradually from a relationship in which one can scarcely distinguish between plant and animal, are perfected antithetically—the plant being ultimately glorified, fixed and rigid, in the tree, and the animal, with utmost mobility and freedom, in mankind." Again, in an essay on some fossil bones of an extinct species of ox: "In any case this ancient creature may be considered a widely distributed parent stock of which the common ox and zebu may be looked upon as descendants." [From "Fossiler Stier," in *Goethes Werke* (Leipzig und Wien: Bibliographisches Institut, n.d.), XXIX, "Schriften zur Naturwissenschaft," I, 430.] It is difficult to avoid drawing the conclusion from such statements that Goethe was a pre-Darwinian. Lamarck's theory had now been published (1809), but Goethe was not aware of it.

necessarily to rest on determining the point in time (history) at which one may say of a concept: "This is it, full-blown." With reference to those who propose Goethe as a pre-Darwinian, some say that after Columbus it is easy to stand an egg on end. It is better, perhaps, to suggest that it detracts not a whit from Columbus' fame that explorers of lesser reknown reached America before him. So it is with Darwin and Goethe. Indeed, even Empedocles thought, on philosophical grounds, that development of organisms is a gradual process wherein imperfect forms are in time replaced by more perfect ones!

And the idea of the *Urphenomen* is not dead. Spengler saw it in history:

"Culture is the archetypal phenomenon of all past and future world history. The profound and little-appreciated idea of Goethe, which he discovered in his 'living nature' and which always formed the base of his morphological research, is to be applied here in its strictest sense to all civilizations in human history, whether completely matured, cut off in full flower, half-grown, or nipped in the bud. It is a method of perception, rather than of analysis. 'The highest that man can attain to is wonder, and if the archetypal phenomenon causes him to wonder, let him be content; nothing higher can offer it to him, and nothing further should be sought behind it; here the limit is drawn.' The archetypal phenomenon is that which represents the pure essence of growth.... It was an insight into things such as Leibniz might have understood. The century of Darwin has remained as remote as possible from it."[48]

Working continuously almost to the end of his life to complete *Faust*, Goethe nevertheless maintained his lively interest in science; witness his concern over the debate in the French Academy between Cuvier and St. Hilaire in 1830. He maintained in these last years an intense curiosity concerning new events and ideas, and conversed on the possibility of connecting the Gulf of Mexico with the Pacific Ocean, on the flying machine, on steam ships, the Suez Canal, and other seemingly fanciful topics.

During the last year of his life, at the age of 82, Goethe published an essay on the spiral tendency in plants, a subject introduced two years previously in a series of lectures by Karl Friedrich Philipp von Martius, a professor in the University of Munich; a second essay on the subject was published posthumously. In these essays, the growth of plants is considered to be the result of two tendencies: the vertical tendency responsible for elongation of root and stem, and the spiral tendency responsible for the distribution of leaves and flower parts on the stem. Goethe adopted

---

[48] Spengler, *op. cit.*, pp. 140–141.

this concept, extended it, and made it compatible with his theory of metamorphosis. Fascinated by the spiral vessels, Goethe devoted considerable attention to them. The reader should enjoy Goethe's accounts of the spiral growth of specially chosen plants, notably the twiners.[49] This subject of spiral growth was thoroughly treated by Darwin, who gave it the name *circumnutation*.[50]

The epoch that was Goethe's life ended on the morning of March 22, 1832. It had marked not only the zenith of German achievement in literature and music, but also the beginning of the era of modern science. Comparative morphology had become an established science. Dalton had defined the atom; Fraunhofer had detected the dark lines in the sun's spectrum. Six years after Goethe's death the world was given the cell theory, a theory Goethe would undoubtedly have received with enthusiasm. The two greatest concepts of science were shortly to come—Helmholz' law of conservation of energy in 1847, and Darwin's principle of organic evolution in 1859.

We have dealt with only one aspect of the many-faceted life of a great man. Many questions remain unanswered. Whether he was a great biologist the reader may judge for himself. Whether his science interfered with his poetry can hardly be judged at all. It is quite possible, as Troll has suggested, that the focal point of his entire mental life is to be found in his scientific writings. Yet, no matter how much our opinions may be at variance, the primary fact remains that Goethe's fundamental and original thoughts as described in his morphology are provocative in the field of biology to this day.[51]

CHARLES J. ENGARD

University of Hawaii
August, 1951

---

[49] See pp. 139–140.
[50] C. Darwin, *The Power of Movement in Plants* (New York: D. Appleton & Co., 1897), p. 1.
[51] Dr. Engard's efforts to complete the introduction were interrupted at this point by his untimely death. The last paragraph was left unfinished.

On Morphology

PARABASIS

Joyfully deliberating,
Striving in the years long past
To understand how, in creating,
Nature lives—I saw at last
Ever the One itself revealing
As myriad ephemera,
Plurality the One concealing,
Governed each by inner law.

Everlasting, evanescent,
Inaccessible, yet near,
Formed, transformed, through change incessant—
Clothed in wonder I am here!

—GOETHE

(English rendering by Aldyth Morris)

Formation and Transformation[1]

> Lo, He passes by me
> Before I am aware of it,
> And is transformed
> Before I can take note of it.
>
> Job[2]

Our Undertaking Is Defended

WHEN A MAN of lively intellect first responds to Nature's challenge to be understood, he feels irresistibly tempted to impose his will upon the natural objects he is studying. Before long, however, they close in upon him with such force as to make him realize that he in turn must now acknowledge their might and hold in respect the authority they exert over him. Hardly is he convinced of this reciprocal influence when he becomes aware of a twofold infinitude: in the natural objects, of the diversity of life and growth and of vitally interlocking relationships; in himself, of the possibility of endless development through always keeping his mind receptive and disciplining it in new forms of assimilation and procedure.

---

[1] This essay comprised the introduction to the first issue of a series of notes and essays which Goethe published in journal form from 1817 to 1824, *Natural Science in General; Morphology in Particular,* from which most of the material translated in this volume has been taken, specifically from the section entitled "Formation and Transformation of Organic Natures." The date, 1807, of the first two sections of the essay shows how long Goethe had had the journal project in mind before actually carrying it out. See pp. 255–257 for the journal's table of contents.

[2] Goethe used Luther's translation of the passage, which has since been officially corrected in the Lutheran Bible and now conforms in meaning with the St. James version:

> Lo, He goeth by me,
> And I see Him not;
> He passeth on also,
> But I perceive Him not.

Luther's version contains the core of Goethe's morphological thinking, namely, that each organic formation is a chain of transformations, and that operating in each organic nature, each manifestation of life, are two principles: the type, working internally, and the influence of environment, working externally. In the passage from Job, Goethe interprets the "He" as the active God-Nature as revealed in the development of organic life.

These circumstances give high pleasure and would insure one's happiness in life if obstacles from within and without did not block the beautiful path to perfection. The years, which at first gave freely, now begin to take their toll; we are satisfied, each in his own measure, with what has been accomplished, and we are quite content to rejoice over it in private, inasmuch as genuine, pure, and stimulating sympathy on the part of others is rare.

How few are inspired by what is perceptible to the intellect alone! The senses, the emotions, the soul exert far greater power over us, and rightly so, since we are destined for active life and not for meditation.

Alas, even in those devoted to understanding and knowledge we seldom find the interest we hope for! The pragmatist—taking note of the special case, observing and analyzing in detail—is inclined to regard as an encumbrance those things that stem from and lead back again to an idea. In his way he feels at home in his labyrinth and takes no interest in a thread that would guide him, not only more rapidly but all the way through. To such a man, metal that is unminted and cannot be counted is a burdensome possession. On the other hand, a man of wider horizons is all too inclined to disdain detail and to reduce to a deadening generality what possesses vitality only as a particular.

In this conflict we have long been engaged. Much has been accomplished in the process, much has been destroyed; and I should never be tempted to entrust my views on Nature to the seas of opinion, in so weak a craft, if we had not had occasion in recent hours of danger[3] to feel all too keenly the value of papers on which we have felt impelled to set down a part of our being.

Let, therefore, what in dauntless youth I had often dreamed of as a finished work, now go forth as a sketch, as a mere fragmentary collection, to function and serve for what it is.

This much it was necessary for me to say in commending these early sketches—some individual parts of which, however, are more or less complete—to the good will of my contemporaries. A good many things that may still remain to be said, we had best introduce in the course of our undertaking.—*Jena, 1807.*

Our Objective Is Stated

When our interest in natural objects, especially the organic ones, is awakened to the extent that we desire to obtain an insight into relation-

---

[3] The plundering of Weimar by the French in October, 1806, had endangered Goethe's papers and collections.

ships between character and function, we believe ourselves best able to acquire such knowledge through analysis of the parts. This method is indeed likely to take us far—it requires but a word or two to remind friends of science what chemistry and anatomy have contributed toward an intensive and extensive view of Nature.

But these analytical efforts, if continued indefinitely, have their disadvantages. To be sure, the living thing is separated into its elements, but one cannot put these elements together again and give them life. This is true even of inorganic bodies, to say nothing of organic ones.

For this reason, the man of science has always evinced a tendency to recognize living forms as such, to understand their outwardly visible and tangible parts in relation to one another, to lay hold of them as indicia of the inner parts, and thus, in contemplation,[4] to acquire a degree of mastery over the whole. How closely this scientific aspiration is bound up with the creative and imitative urges need not be dealt with in detail.

Hence several attempts are found in the progress of art, learning, and science to establish and develop a theory to which we should like to give the name "morphology." The varied forms these attempts assume, will be spoken of in the historical portion of our work.

The German language has the word "Gestalt"[5] to designate the complex of life in an actual organism. In this expression the element of mutability is left out of consideration: it is assumed that whatever forms a composite whole is made fast, is cut off, and is fixed in its character.

However, when we study forms, the organic ones in particular, nowhere do we find permanence, repose, or termination. We find rather that everything is in ceaseless flux. This is why our language makes such frequent use of the term "Bildung" to designate what has been brought forth and likewise what is in the process of being brought forth.

In introducing a science of morphology, we must avoid speaking in terms of what is fixed. Thus, if we use the term "Gestalt" at all, we

---

[4] Goethe's method of thinking has been described as contemplative cognition, "a mode of cognition which is at once sensory (grasping the phenomenon) and spiritual (perceiving the spirit which manifests itself in the phenomenon)." Karl Vietor, *Goethe the Thinker* (Cambridge: Harvard University Press, 1950), p. 12. For discussion of Goethe's methods by contemporaries, see pp. 191–195, 235 of this translation.

[5] Goethe's discussion in the following paragraphs revolves about the derivation of the nouns, *Gestalt* and *Bildung*. *Gestalt* is derived from the Middle High German past participle of *stellen: to set, place, put; Bildung* is derived from the verb *bilden*, allied in meaning to, though not a translation of, the English verb *to build*. The distinction between the two German nouns may be brought out by translating *Gestalt* as *form*, *Bildung* as *formation*.

ought to have in mind only an abstract idea or concept, or something which in actuality is held fast for but an instant.

What has just been formed is instantly transformed, and if we would arrive, to some degree, at a vital intuition of Nature, we must strive to keep ourselves as flexible and pliable as the example she herself provides.

If we divide an organism into its anatomical parts and then in turn divide these parts into their components, we finally come to such beginnings as have been labeled "similar parts." We are not speaking of these here;[6] we are rather pointing out a higher law of the organism, which we shall explain as follows:

Each living creature is a complex, not a unit; even when it appears to be an individual, it nevertheless remains an aggregation of living and independent parts, identical in idea and disposition, but in outward appearance identical or similar, unlike or dissimilar. These organisms are partly united by origin; partly they discover each other and unite. They separate and seek each other out again, thus bringing about endless production in all ways and in all directions.

The more imperfect a creature is, the more do these parts appear identical or similar to each other and the more do they resemble the whole. The more the creature is perfected, the more dissimilar its parts become. In the first case, the whole is more or less identical to the parts; in the second case, the whole is dissimilar to the parts. The more the parts resemble each other, the less they are subordinated to each other. Subordination of the parts betokens a more perfected creature.

Since in all generalizations, be they ever so well thought out, there is something incomprehensible for the person who cannot apply them, who cannot buttress them with the necessary examples, it is our intention to present some examples here at the beginning, but only a few, as our whole book, after all, is devoted to exposition and pursuit of these and other ideas and principles.

That a plant or even a tree, though it appears to us as an individual, consists purely of detached parts resembling both each other and the whole—of this fact there is no doubt. How many plants are indeed propagated by slips! The bud of the least complex variety of fruit tree puts forth a shoot that in turn produces a number of identical buds; and it is precisely in this manner that propagation through seed takes place. Such propagation is the development of countless identical individuals from the womb of the mother plant.

Here one can readily see that the secret of propagation through seeds

---

[6] Thus, not of atoms and molecules.

ON MORPHOLOGY

is already expressed in the principle just discussed, and if one but observes and considers the matter properly, one discovers that even a seed, which to us appears as a unit, is in reality a complex of identical and similar organisms. The bean is usually cited as a clear example of germination. If one takes a bean in its completely enfolded state before it has germinated, upon opening it he will discover first the two seed leaves [Fig. 2]. These cannot legitimately be compared with the placenta, for they are genuine leaves, merely distended and filled out with a meal-like substance, which will turn green in light and air. Further, one already discovers the plumule here, which in turn represents two leaves, more complex and capable of even greater complexity. If one then recalls that a bud reposes behind each leaf stalk, theoretically if not actually, one will recognize in a seemingly simple seed a complex of several units that may be called identical in the abstract though only similar in outward appearance.

What is identical in the abstract may in actuality be either identical or similar, may indeed be even completely unlike or dissimilar—in this fact consists the variable life of Nature that we propose to outline in the following pages.

By way of further introduction, let us cite an example from the lowest level of the animal kingdom. There are infusoria[7] of a rather simple form which are visibly mobile in moisture, but which burst as soon as this moisture has evaporated, ejecting a quantity of the grains into which, had they remained in moisture, they probably would have separated in the natural course of events and thereby have produced an infinite progeny. However, let this much suffice here, as our theory will reappear continually throughout the exposition proper.

If one looks at plants and animals in their most rudimentary stages, they are scarcely distinguishable from one another. Such nuclear organisms—whether rigid, mobile, or semi-mobile—are just barely perceptible to our senses. Whether these first beginnings could be conclusively traced in opposing directions, to the plant through light and to the animal through darkness, I do not make bold to decide, although opinions and analogies are not lacking on this subject. This much we can say: creatures, emerging gradually from a relationship in which one can scarcely distinguish between plant and animal, are perfected antithetically—the plant being ultimately glorified, fixed and rigid, in the tree; and the animal, with utmost mobility and freedom, in mankind.

Gemmation and proliferation are two further laws of the organism

---

[7] Several of Goethe's letters during March and April, 1786, reveal that he was zealously studying infusoria.

which derive from that principle of the coexistence of identical and analogous natures, and which actually express that coexistence in dual fashion only. We shall seek to follow these two paths throughout the organic realm, a process in which much will be graphically resolved into sequences and categories.

If we observe the vegetative type of organic life, a top and bottom are instantly apparent. Occupying the lower position is the root, the action of which is directed earthward and which is associated with moisture and darkness, whereas the stem, trunk, or whatever corresponds to it, strives skyward in the opposite direction, toward the light and the air.

As we now observe this magic structure, and acquire greater insight into the way it lifts itself upward, we again encounter an important principle of organization: that no life can operate on an exposed surface or exercise its reproductive power there; that instead all life activity demands a covering to shield it against the outward rough element, be it water, air, or light, and to safeguard its delicate existence, in order that it may fulfill the specific function of its inner nature. Whether the covering has the appearance of bark, skin, or peel—everything that emerges into life, everything that has a vital function must be enveloped. Thus, too, everything that is turned to the outside is gradually and prematurely subject to death, to decomposition. The bark of trees, the skin of insects, the hair and feathers of animals, even the epidermis of humans, are coverings which are eternally detaching themselves, sloughing off, resigning themselves to death, behind which new coverings are constantly forming, and under which, near the surface or deeper, life weaves its creative webwork.—*Jena, 1807.*

The Content Is Given a Foreword

Of the present collection[8] only the essay on the metamorphosis of plants has previously appeared in print. Published separately in 1790, it met with a cold, almost unfriendly, reception. Such antipathy was quite natural, for the concept of preformation,[9] of successive development of what had already been in existence from the time of Adam onward, had generally obsessed even the best minds. Then, too, Linné—vigorous of intellect, discerning and authoritative—had introduced an

---

[8] *Natural Science in General; Morphology in Particular,* Vol. I, No. 1 (1817).

[9] The theory of preformation, which played a dominant role in the 18th century, held that the organism was contained, fully formed and complete in all its parts, in the germ cell, merely increasing from minute proportions to adult size in the process of development.

expository method, with special reference to plant formation, which was more acceptable to the times.[10]

My earnest efforts thus remained wholly without influence; and, satisfied that I had found the guidepost to my own private path, I merely observed with greater care the relationship and reciprocal effect of normal and abnormal phenomena, taking careful note of everything that experience yielded obligingly and bit by bit. At the same time, I spent an entire summer with a series of experiments designed to teach me how formation of fruit can be prevented through superfluous nourishment and accelerated through curtailed nourishment.

I took advantage of the opportunity of arbitrarily illuminating or darkening a hothouse to acquaint myself with the effect of light on plants; the phenomena of blanching and etiolation occupied me particularly; experiments with colored panes were begun.

When I had acquired sufficient skill to judge the organic formation and transformation of the plant world in most cases and to recognize and trace the sequence of forms, I likewise felt impelled to study in greater detail the metamorphosis of insects.

No one would deny that metamorphosis occurs in the case of insects, for the life course of such creatures is unceasing transformation, visible to the eye and tangible to the hand. My earlier knowledge of the subject, derived from the raising of silkworms over a period of years, was still intact, and I now extended it by observing several genera and species from egg to butterfly. The most valuable of the plates that were made under my direction at that time are still preserved.

In the metamorphosis of insects I found no contradiction with what has been transmitted in treatises, and I needed only to evolve a chart in table form, so that the individual observations could be noted in logical sequence and the wonderful life course of such creatures could be seen in clear perspective.

Of these efforts, too, I shall attempt to give an account—quite dispassionately, as here my view is not opposed to that of anyone else.

During the period of this study I also gave my attention to comparative anatomy of animals, particularly of mammals, a study for which a great interest had already awakened within me. Buffon[11] and Daubenton[12] achieved much. Camper[13] appeared as a shining light in intellect,

---

[10] For discussion of theory of anticipation, see pp. 74–76.
[11] Georges-Louis Leclerc, Count de Buffon, 1707–1788, French naturalist, of great influence on Goethe's scientific views.
[12] Louis Jean Marie Daubenton, 1716–1799, collaborator in Buffon's zoological work.   [13] Petrus Camper, 1722–1789, famous Dutch anatomist and naturalist.

knowledge, talent, and activity; Sömmering[14] proved admirable; Merck[15] turned his ever dynamic efforts to such subjects. With these last three I kept in constant touch—Camper by way of correspondence, and the other two through friendship maintained even during separations.

In the course of our study of physiognomy, the significance of forms and their mutability necessarily occupied our attention by turns; much on this subject had also been discussed and worked out with Lavater.[16]

At a later period, by virtue of long and frequent sojourns in Jena, and Loder's[17] indefatigable gift for teaching, I could very soon take pride in my insight into animal and human development.

The method previously adopted in my study of plants and insects also guided me along this path: for when isolating and comparing their forms, the subject of formation and transformation necessarily came up for discussion.

Those times, however, were darker than it is now possible to imagine. It was asserted, for example, that if man only wished he might be able to walk comfortably on all fours, and that bears might become human beings if they were to hold themselves erect for a while. The intrepid Diderot ventured to make certain proposals for producing goat-footed fawns to be installed in livery, by way of special pomp and distinction, on the coaches of the rich and great.

For a long time it was impossible to discover what it was that distinguished humans from animals. It was believed that one could definitely distinguish the ape from man by the fact that apes carried their four incisors in a bone,[18] the presence of which could actually be verified empirically. Thus all knowledge fluctuated, seriously and in jest, between attempts to verify the half-true and to lend luster of a sort to the false, all the while maintaining an arbitrary, capricious activity. The greatest confusion, however, resulted from the argument as to whether beauty was something actual and inherent in objects or something relative, con-

---

[14] Samuel Thomas von Sömmering, 1755–1830, German anatomist and physiologist.

[15] Johann Heinrich Merck, 1741–1791, writer, editor, outstanding paleontologist, Goethe's friend.

[16] Johann Kaspar Lavater, 1741–1801, Swiss theologian and physiognomist. It was collaboration in Lavater's project, *Physiognomische Fragmente*, in 1774, that first stimulated Goethe to independent research in natural science.

[17] Justus Christian Loder, 1753–1832; as teacher of anatomy in Jena, Loder was of decisive influence on Goethe at the time the latter began his biological studies.

[18] The intermaxillary. Its discovery in man, by Goethe in 1784, invalidated the belief that man was separated from even the higher animals through the lack of this bone. See p. 5.

ventional, indeed individual, to be ascribed to the beholder and connoisseur.

In the meantime, I had devoted myself exclusively to osteology, for it is in the skeleton that the decisive character of each form is safely and forever preserved for us. I gathered the remains of bones about me, some old and some new, and in my travels I delved about in museums and private collections for creatures with formations that might be illuminating to me, as a whole or in individual parts.

In doing this, I soon felt the necessity of establishing a type, against which one might gauge all mammals for conformity and deviation; and just as I had once sought out the archetypal plant, I now sought to find the archetypal animal.

My labored, tormented research was lightened, indeed sweetened, when Herder[19] undertook to write down his *Ideas on the History of Humanity*. Our daily conversations were concerned with the prime origins of the water-earth[20] and the organic creatures developing on it from time untold. This unceasing development from primordial beginnings was continually under discussion, and our scientific knowledge was clarified and enriched day by day through this sharing and countering of ideas.

With other friends[21] I likewise had lively conversations on these subjects of passionate interest to me, and such discussions were not without influence and mutual benefit. Indeed, it would perhaps not be presumptuous to imagine that a good deal that sprang into being in that way and was propagated in the scientific world by word of mouth is now bearing fruit that we delight in, even though we cannot always call by name the garden that produced the graft shoot.

At present, by virtue of ever-widening experience and ever-deepening philosophy, much has become accessible that was not available to me and others when the following pages were written. Should they seem superfluous, let their content be looked upon from the historical viewpoint, as a chronicle of quiet, steadfast, and consistent effort.

---

[19] Johann Gottfried Herder, 1744–1803, theologian, philosopher, poet, critic; Goethe's friend since 1771, and of decisive influence in the development of Goethe; called to Weimar in 1776 as court chaplain, through the instrumentality of Goethe. The work mentioned here appeared from 1784–1791.

[20] The concept of Buffon, whose theory it was that the earth had once been completely or almost completely covered with water and that the first forms of life had appeared at this period.

[21] Alexander von Humboldt, for one. See pp. 122, 183 for other references to Humboldt.

Tarassei tous anthropous ou ta pragmata, alla ta peri ton pragmaton dogmata.

What disturbs men's minds is not things themselves but the interpretations placed upon them.

        Epictetus, *Manual*

Non quidem me fugit nebulis subinde hoc emersuris iter offundi, istae tamen dissipabunter facile, ubi plurimum uti licebit experimentorum luce; natura enim sibi semper est similis, licet nobis saepe ob necessariarum defectum observationum a se dissentire videatur.

I am quite aware that this road is obscured by mists that may pass over it from time to time. Yet these mists will be easily dispersed as soon as it is possible to employ widely the light of experiments. For Nature remains always the same; when she seems to be different it is because of the inevitable defect of our observations.

        Linné, *Prolepsis of Plants*

The Metamorphosis of Plants[22]

Introduction

1. Anyone who devotes the least attention to the growth of plants can easily note that certain of their external parts are often transformed, assuming, either completely or to some lesser degree, the form of neighboring parts.

2. For example, it is usually by developing additional petals instead of filaments and anthers that a single flower is transformed into a double one. Such petals are then either identical in form and color with the other leaves of the corolla, or may still bear visible traces of their origin.

3. Once we observe that it is possible in this way for the plant to take a step backward and reverse the order of growth, we become all the more alert to Nature's regular procedure and become familiar with the laws of transformation by which she brings forth one part through another, achieving the most diversified forms through modification of a single organ.

4. The intimate relationship of various external plant parts—such as leaves, calyx, corolla, and stamens—which develop one after another, and apparently from one another, has long been recognized by naturalists in a general way. Indeed, it has even been studied in detail, and the process by which one and the same organ makes its appearance in multifarious forms has been named the *metamorphosis of plants*.

5. This metamorphosis may be of three different types: *regular*, *irregular*, and *accidental*.

6. *Regular* metamorphosis we might also call *progressive*, for it is this type that may be observed at work step by step from the first seed leaves to the final development of the fruit. By transmutation of one form into another, it ascends as though on the rungs of an imaginary ladder to that climax of Nature, reproduction through two sexes. It is this type of metamorphosis which I have been studying attentively for several years and now undertake to explain in this essay. In the following demonstration, we shall therefore consider the plant only insofar as it is an annual, advancing continuously from seed to fructification [Fig. 1].

7. *Irregular* metamorphosis we might equally well term *retrogressive*. Whereas in regular metamorphosis Nature hurries forward to her goal,

---

[22] See pp. 165–167 for Goethe's account of the genesis and history of this essay.

here she takes a step or two backward. In the one case, she fashions the flowers and prepares them for their acts of love with irresistible urge and powerful effort; in the other, she slackens, so to speak, and irresolutely leaves her creations in a soft and indeterminate state which is often pleasing to the eye, yet internally impotent and ineffectual. Through the knowledge we gain from studying this latter type of metamorphosis, we shall be able to bring to light what the regular type keeps hidden from view and to distinguish clearly what otherwise we are allowed only to conjecture. It is by this procedure that we have the best prospect of attaining our purpose.

8. On the other hand, we shall disregard the third type of metamorphosis, which is effected accidentally by outside agents, chiefly insects, since it might divert us from the single path marked out for us and thus defeat our purpose. Perhaps there will be an opportunity elsewhere to speak of those excrescences, which are abnormal and yet restricted within definite limits.

9. I have ventured to prepare this essay without supplementary illustrative plates, though from some points of view they might seem necessary. It is my intention to reserve them for a sequel, a plan which is quite feasible, inasmuch as enough subject matter remains to clarify and elaborate this short and merely preliminary treatise. In the sequel it will not be necessary to maintain the present formal pace, making it possible to introduce related material and to find an appropriate place for several passages from authors whose views are similar to my own. Especially, I will not fail to make use of all suggestions from contemporary experts engaged in this noble science. It is to them that I now submit and dedicate these pages.

FIGURE 1

Annual, representative of basic plant in Goethe's discussion of metamorphosis; flower parts, separated for purpose of illustration, from top to bottom; pistil, stamens, corolla, and calyx.

ON MORPHOLOGY

I. *Concerning the Seed Leaves*

10. Since we have undertaken to observe the sequence of plant growth, we shall straightway direct our attention to the moment when the plant emerges from the seed. In this phase we can recognize its immediate parts with ease and accuracy, for the seed coats, which indeed do not concern us here, have more or less been cast off within the earth;

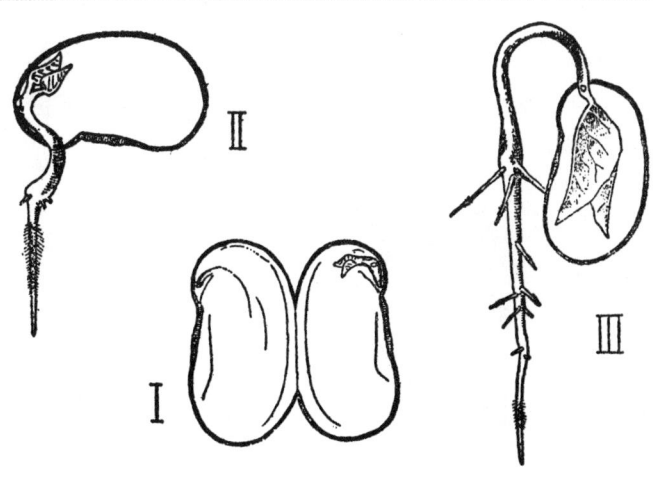

FIGURE 2

Germination of garden bean. I. Seed opened, cotyledons separated to reveal embryo. II. Bean in process of germination, one cotyledon removed to reveal growing embryo, the latter now with strong root and radically increased in size. III. Advanced stage of germination.

and in many cases, when the root has fastened itself into the soil, the plant brings into full view the primary organs of its upper growth, which up until now have been hidden within the seed coat.

11. These primary organs are known by the name *cotyledons*. Botanists have also named them seed valves, kernel pieces, seed lobes, and seed leaves, attempting thus to designate the various forms in which we perceive them.

12. They are often misshapen, crammed, as it were, with crude matter, and as much expanded in thickness as in breadth; their vessels are unrecognizable and scarcely distinguishable from the mass as a whole. They bear almost no resemblance to a leaf, and we might be misled into regarding them as special organs.

13. Yet in many plants these cotyledons approach leaf form: they flatten out; exposed to light and air, they assume a deeper shade of green; their vessels become distinct and begin to resemble veins.

14. Finally they appear before us as true leaves: their vessels are capable of the finest development; their similarity to the subsequent leaves

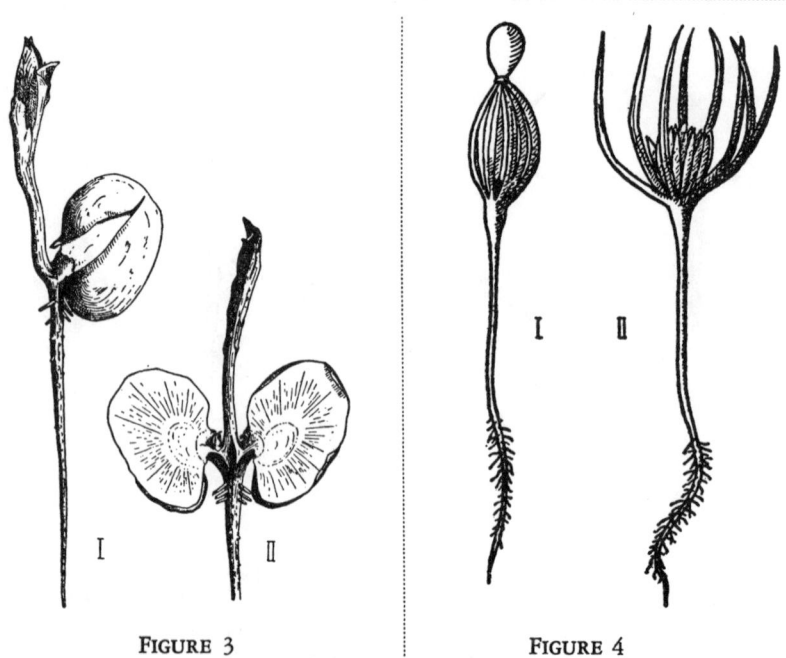

FIGURE 3   FIGURE 4

FIG. 3—Young seedling of *Vicia Faba*. I. Seed coat burst open; seedling developing strong root and young shoot. II. Cotyledons separated, with "eye," or bud, in each axil. FIG. 4—Young seedling of *Pinus maritima*; cotyledons collected around an axis (after de Candolle). I. Tips of cotyledons still held in seed coat. II. Seed coat cast off.

will not permit us to consider them separate organs; and we recognize them instead as the first leaves of the stem.

15. But if a leaf is inconceivable without a node, or a node without a bud, we may conclude that the point where the cotyledons are attached is the first true nodal point of the plant [Fig. 13]. This view is borne out by those plants which produce young buds at the immediate base of the cotyledonary wings, and develop complete branches from these first nodes, as *Vicia Faba* customarily does [Fig. 3].

Seedling of bean (center) and of Turkish maize (left and right).
Original done in color under Goethe's supervision.

PLATE I

Sunflower, showing how "stem leaves gradually contract, change, and gently creep . . . into the calyx."
From *Hortus Eystettensis*.

PLATE II

ON MORPHOLOGY 37

16. The cotyledons are usually double [Plate I], and, in this connection, we must point out what will later seem still more important, namely, that the leaves of the first node are often *paired,* even when the succeeding leaves stand *alternately* on the stem. We are witnessing here an approach and union of parts that Nature later disjoins and separates. This is even more striking when cotyledons appear in the form of numerous leaflets clustered about a single axis, while the stem, gradually developing from its midst, produces its leaves individually, round about itself. Such a case may be observed very clearly in the growth of conifers, where a wreath of needles forms a kind of calyx [Fig. 4]. We shall later be obliged to recall this present instance in discussing similar phenomena.

17. Isolated cases of extremely irregular cotyledons of plants that germinate with only one leaf, we shall pass over for the present.

18. On the other hand, we point out that even the most leaflike cotyledons are always rather undeveloped as compared with the subsequent leaves of the stem. Their outlines, in particular, are of utmost simplicity, and exhibit as little trace of indentations as their surfaces do of hair or other vessels typical of mature leaves.

II. *Development of Stem Leaves from Node to Node*

19. We can henceforward observe the successive development of the leaves in detail, for now the progressive operations of Nature all take place before our very eyes. Often a number of the succeeding leaves are already present in the seed, locked within the cotyledons and known as plumules in their folded state [Fig. 2]. The relationship of their form to that of the cotyledons and the following leaves differs among various plants, but, for the most part, they already deviate from the cotyledons in that their form is flat, delicate, and generally that of true leaves: they are entirely green in color; they rest on a visible node; and they bear an undeniable relationship to the following stem leaves, although they are still inferior to them in that their periphery or margins are not yet fully elaborated.

20. Yet development continues irresistibly from node to node throughout the leaf, as the midrib elongates and the secondary ribs arising from it reach out more or less toward the sides. These variations in relationship between the different ribs are the chief reason for the diversification of the leaf forms [Figs. 5, 6, 9]. The leaves are by now notched, deeply indented, or composed of several leaflets, having, in the latter case,

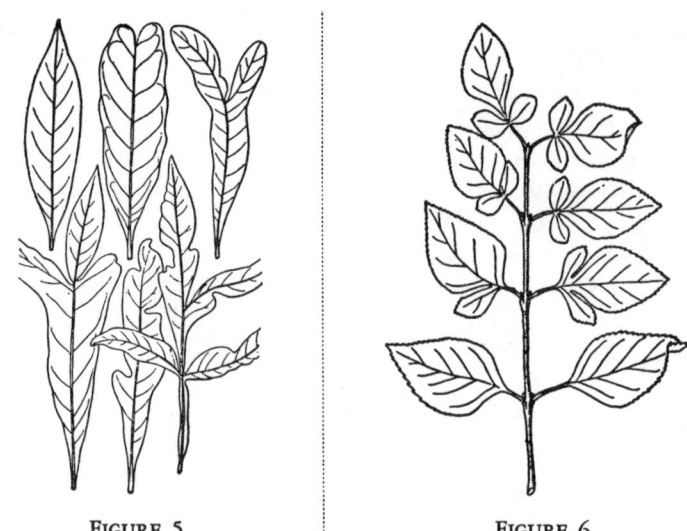

FIGURE 5   FIGURE 6

FIG. 5—Various leaf shapes of *Sapindus saponaria*, showing transformation of a simple leaf into a pinnately compound one (after de Candolle). FIG. 6—Twig of *Forsythia*, showing simple to pinnately compound leaves.

the appearance of complete little branches. Of such successive and extreme diversification of the simplest leaf form, the date palm affords a striking example [Fig. 7]. In a succession of several leaves, the midrib thrusts itself forward; the fanlike, simple leaf is torn and divided; and a highly elaborate leaf develops, vying with a branch in complexity.

21. The leaf stalk develops in direct proportion to the increasing complexity of the leaf, regardless of whether it be directly connected with its leaf, or form an independent stalklet which later may be easily detached.

22. Such an independent leaf stalk has likewise an inclination to transform itself into leaf form, as we may observe in various plants, for example, the orange tribe [Fig. 8]. The organization of such stalks will require further discussion in the sequel, but for the present we shall pass over the subject.

23. Nor will we enter at this time into detailed observation of the stipules; we shall merely remark, in passing, that they also are remarkably transformed during the subsequent alteration of the leaf stalk, especially when they constitute a part of it [Fig. 10].

ON MORPHOLOGY 39

24. Whereas the leaves are indebted for their first nourishment principally to the more or less modified watery elements they derive from the stem, their major development and complexity they owe to light and air. We have found that the cotyledons, which are produced in the enclosed seed coat and are filled to the brim, as it were, with a very crude sap, are scarcely organized and developed at all, or at best roughly so. Now we likewise find that the leaves of underwater plants are more crudely organized than those of plants exposed to the open air. Indeed, the same plant species will even develop smoother and less complex leaves when it grows in low, moist places, and rough, hairy, more delicately elaborated leaves when transferred to higher regions.

25. In the same way, purer types of gas will at least greatly foster

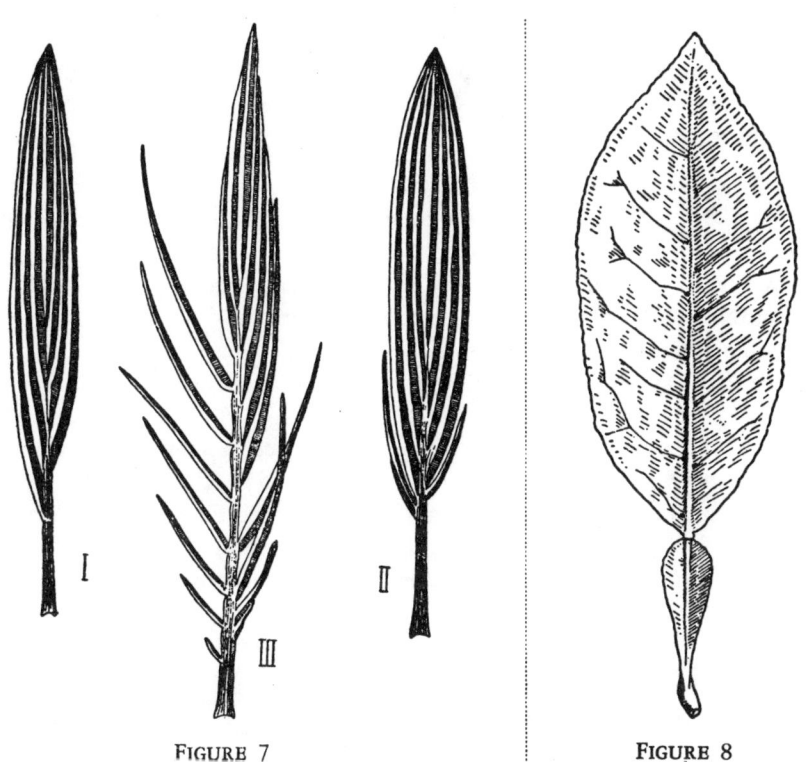

FIGURE 7   FIGURE 8

FIG. 7—Leaf development of the date palm, *Phoenix dactylifera*. I. Simple first leaf. II. Beginning of articulation. III. Advanced stage of articulation.
FIG. 8—Leaf of orange, *Citrus*, with alate petiole.

anastomosis of the vessels if not actually bring it about completely. These vessels originate from the ribs and meet to form the skin of the leaf, much as though their tips had set out in search of one another. We are inclined to attribute the threadlike or antler-like form of many underwater plants to lack of complete anastomosis. A clear and instructive example is afforded in the growth of *Ranunculus aquaticus,* for those of its leaves that are produced under water consist of threadlike ribs, whereas the ones that have developed above water have been completely anastomosed and formed into a connected surface [Fig. 11]. Indeed, the transition may be clearly noted in half-anastomosed, half-threadlike leaves of this plant [Fig. 12].

26. By experimentation it has been learned that leaves absorb various types of gas, combining them with their internal moistures; and apparently there is no doubt that they send these purer saps back again to the stem, thereby fostering remarkably the development of the nearby buds. Examination in several plants of the types of gas developed from the leaves, and even from the hollow parts of the reed, has established conclusive proof.

27. We observe in some plants that one node develops from another. This is strikingly evident in stems that are closed from node to node, such as those of cereals, grasses, and reeds; it is less obvious in those plants whose stems are open throughout and are filled with a kind of pith, or rather a cellular tissue. But the important place—in relation to other interior plant parts—which was once assigned to the plant part formerly called marrow, has been disputed, on excellent grounds it seems to us.* The influence it appears to exert on growth has been denied, and all vegetative and reproductive power has been unhesitatingly attributed instead to the inner face of the second rind, the so-called *liber.* Thus it seems all the more convincing to us here that an upper node, originating as it does from the preceding one and obtaining its saps directly through it, must necessarily receive these saps in a purer and more filtered state and must benefit by the previous action of the leaves; also that the node must develop more perfectly and supply its own leaves and buds with more refined juices.

28. While the cruder fluids are in this manner continually drained off and replaced by purer ones, the plant, step by step, achieves the status prescribed by Nature. We see the leaves finally reach their fullest expan-

---

*Hedwig, in the third number of the *Leipzig Magazine.* [Johann Hedwig, 1730–1799, professor of medicine and botany in Leipzig.]

ON MORPHOLOGY

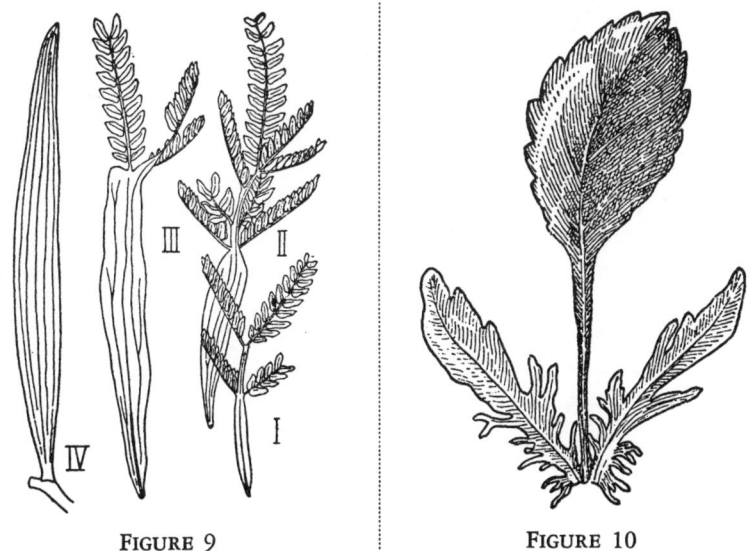

FIGURE 9   FIGURE 10

FIG. 9—Leaves of *Acacia heterophylla*, showing transitions from pinnate leaves to phyllodes. FIG. 10—Leaf of pansy, *Viola tricolor*, with strongly developed stipules.

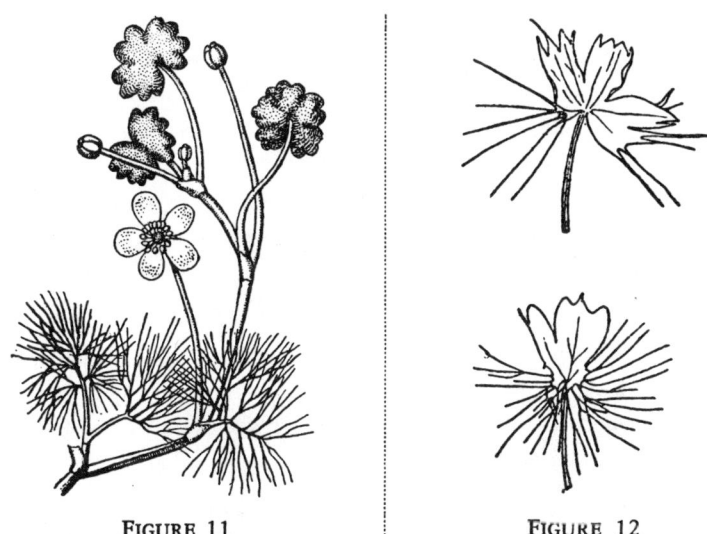

FIGURE 11   FIGURE 12

FIG. 11—*Ranunculus aquaticus*. FIG. 12—Two floating leaves of *Ranunculus aquaticus*, representing transitions to submerged leaves.

sion and elaboration, and soon thereafter we become aware of a new aspect, apprising us that the epoch we have been studying has drawn to a close and that a second is approaching—the epoch of the flower.

III. *Transition to Inflorescence*

29. The transition to the flowering phase is relatively rapid or relatively slow. In the latter case, we usually notice that the stem leaves again begin to draw inward from the periphery and, especially, to lose their diverse marginal divisions. On the other hand, the leaves again begin to expand somewhat in the lower parts, where they are joined to the stem. At the same time, we see the stem assuming a much more delicate and slender form as compared with its former state, even where the interval from node to node is not markedly increased.

30. It has been noted that abundant nutriment retards the flowering of a plant and that moderate, or indeed scanty, nutriment hastens it. Here we see even more clearly the influence of the stem leaves discussed previously. As long as cruder saps remain in the plant, all possible plant organs are compelled to become instruments for draining them off. If excessive nutriment forces its way in, the draining operation must be repeated again and again, rendering inflorescence almost impossible. If the plant is deprived of nourishment, this operation of Nature is facilitated and curtailed. The nodal organs become more refined, the action of the unadulterated saps becomes purer and stronger, and the transformation of the parts is made possible and proceeds irresistibly.

IV. *Formation of the Calyx*

31. When this transformation takes place swiftly, as it often does, the stem is suddenly lengthened and attenuated, pushing upward from the node of the last-developed leaf and collecting several leaves around a center at its tip [Figs. 1, 13].

32. It can be proved most clearly, it seems to us, that the leaves of the calyx are the very same organs which have previously made their appearance as stem leaves, but are now clustered around a common center, often in greatly altered form.

33. In discussing the cotyledons, we have already noted a similar process of Nature, and have seen several leaves, and thus clearly several nodes, collected and arranged side by side around a single point. When conifers emerge from the seed, they exhibit a radiate crown of unmistak-

ON MORPHOLOGY

able needles, which, in contrast with the usual cotyledons, are already quite highly developed [Fig. 4]. Thus, in the early infancy of the plant, we already see a foreshadowing, as it were, of the natural force by which the inflorescence and infructescence will be effected at a more advanced age.

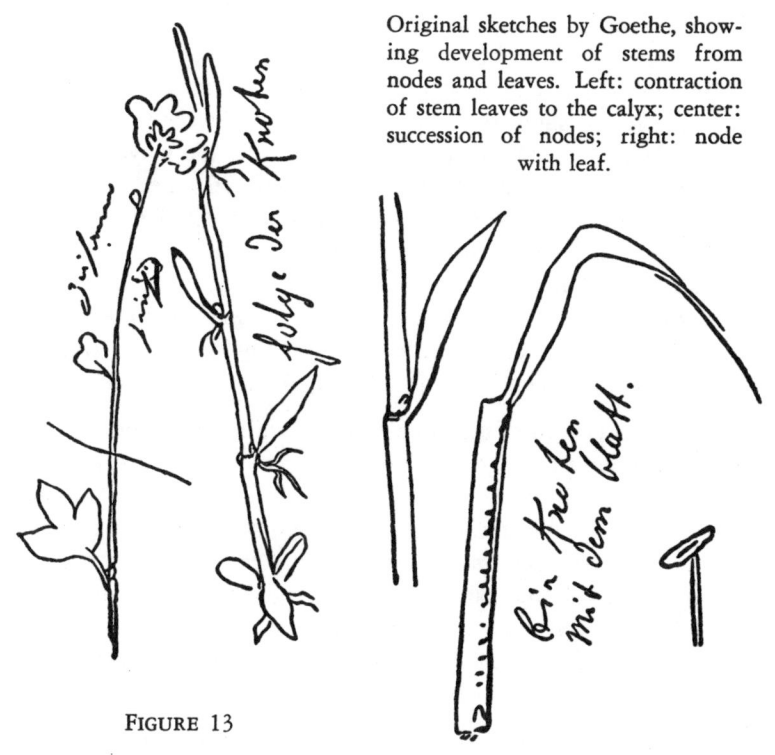

Original sketches by Goethe, showing development of stems from nodes and leaves. Left: contraction of stem leaves to the calyx; center: succession of nodes; right: node with leaf.

FIGURE 13

34. Furthermore, unaltered stem leaves in several flowers are drawn together into a kind of calyx directly below the corolla [Fig. 14]. Since they retain their original leaf form completely, we need no further proof that they are leaves beyond the evidence of our own eyes or the name that botanical terminology has given them, blossom leaves or *Folia floralia*.

35. We shall have to examine rather closely a case already cited, in which the transition to the inflorescence takes place slowly. Here the stem leaves gradually contract, change, and gently creep, so to speak,

into the calyx. Such a process we can very easily observe in the calyxes of composite flowers, especially sunflowers and marigolds [Plate II, Fig. 45].

36. The natural force that collects several leaves around a center brings about an even closer affiliation, and renders these collected and modified leaves still less recognizable by joining them, either completely or partially, and attaching them along the sides. Such tightly pressed and crowded leaves are in close contact and in their delicate condition are anastomosed through the influence of the extremely refined saps by now present in the plant; they acquire bell-shaped, or so-called *one-leaved*, calyxes that clearly show their composite origin by the fact that they are more or less notched and incised along the top. We may have visible evidence of this by comparing a number of deeply indented calyxes with multi-leaved ones and, especially, by examining in detail the calyxes of various composites. For example, we shall see that the calyx of the marigold, which in systematic classification is set down as *simple* and *multipartite,* actually consists of several concrescent and superposed leaves, toward which the contracted stem leaves stealthily make their way, so to speak, in the manner mentioned above [Fig. 45].

37. In many plants, the number and arrangement of the sepals around the stalk as an axis, either individually or fused, remain constant; this is true also of the succeeding members. It is largely this constancy which has made possible the progress of botanical science, and which is responsible for its reputation for accuracy and its growing prestige. In other plants, the number and formation of these parts is not equally constant; but even this absence of constancy did not confuse sharp-sighted experts in the field, for by exact classification they have sought to confine these deviations of Nature, as it were, within a narrower orbit.

38. In this way, then, Nature fashions the calyx: around a single center she *connects* several leaves, and consequently several nodes, which she ordinarily would have produced successively and at some distance *apart,* usually in a certain number and order [Fig. 13]. If the flowering were retarded by the infiltration of superfluous nutriment, the leaves would be separated and would assume their original shape. Thus, in the calyx, Nature forms no new organ but merely combines and modifies organs already known to us, in this way advancing one step nearer her goal.

v. *Formation of the Corolla*

39. We have seen that the calyx originates by means of elaborated saps which are gradually generated in the plant; and now it is again the

ON MORPHOLOGY 45

calyx which is destined to become the organ of further refinement. Indeed, this conception seems plausible even if we explain its operation on a purely mechanical basis. For how exceedingly delicate and capable of the most refined filtration must vessels inevitably become when, as we have seen above, they are closely crowded and pressed against each other.

40. The transition from calyx to corolla we are able to observe in more than one case; for although ordinarily the color of the calyx is still green, resembling the color of the stem leaves, it nevertheless often

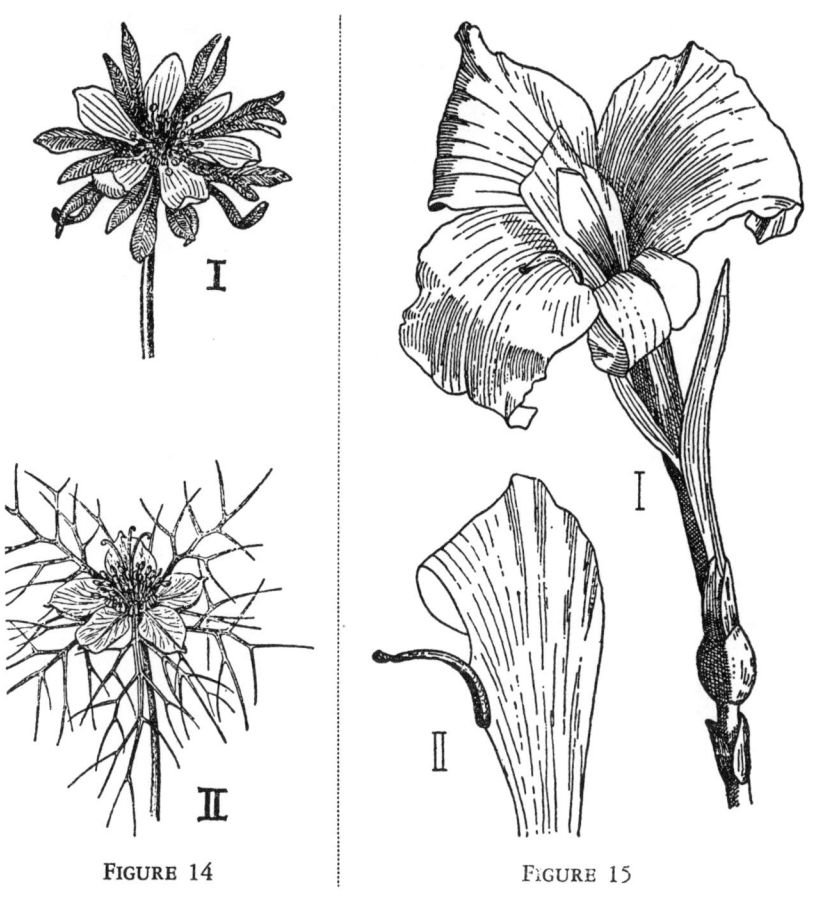

FIGURE 14   FIGURE 15

FIG. 14—"Stem leaves . . . drawn together into a kind of calyx directly below the corolla." I. *Eranthis hiemalis.* II. *Nigella damascena* (after Schönichen).
FIG. 15—*Canna iridiflora.* I. Complete view of flower. II. Individual petal with anther, the petal assuming the role of filament.

changes in one or another of its parts—at the tips, margins, back, or even over its inner surface while the outer surface still remains green; and always we see a refinement associated with this coloration. In this way, ambiguous calyxes arise that might with equal justification be considered corollas.

41. We have already noticed that from the cotyledons onward there is great enlargement and elaboration of the leaf, especially of its periphery, and that thence to the calyx a contraction of the outline occurs. Now we observe that the corolla, likewise, is produced through an expansion. The petals are usually larger than the sepals; refined to a high degree by the influence of purer saps which have been refiltered through the calyx, these same petals that contracted as organs of the calyx can now be observed to expand again and to represent new and entirely different organs. Their delicate organization, their color, and their scent would quite obscure their origin for us if we could not play the spy on Nature in several exceptional cases.

42. For example, within the calyx of a pink there is often a second calyx, which is sometimes fully green and shows a tendency toward a single-leaved, incised calyx, and which, at other times, is jagged and transformed at its tips and margins into actual rudimentary petals—delicate, expanded, and colored. Once again, then, we clearly recognize the relationship of corolla to calyx.

43. The relationship of corolla to stem leaves is disclosed to us in more than one manner, for in several plants the stem leaves are more or less tinted long before they approach the inflorescence, while in others they are completely colored in its vicinity.

44. Sometimes Nature also proceeds directly to the corolla, ignoring the calyx, as it were. In this event, likewise, we have opportunity to observe that stem leaves are transformed into petals [Plate III]. For example, an almost fully developed and colored petal will sometimes appear on tulip stems. And even more remarkable is the case when such a leaf-petal is half green, and is torn into two parts, the green half which is related to the stem remaining attached to it, and the colored part being lifted up with the corolla.

45. It seems quite likely that the color and scent of petals should be attributed to the presence of the male sperm they contain. Apparently, the sperm is not yet sufficiently isolated but is still intermingled and diluted with their other fluids; and the beauty of the colors leads us to think that the material with which the leaves are filled, though already

of a high degree of purity, has not yet reached the highest degree, in which it appears white and untinted.

VI. *Formation of Staminal Organs*

46. All this appears even more credible when we consider the close relationship of petals and staminal organs. If the kinship of all other parts to each other were equally obvious, so universally observed and settled beyond all dispute, the present essay might well be considered superfluous.

47. Nature sometimes shows us normal cases of this transition, for example, in the canna and several other plants of that family. Here, a

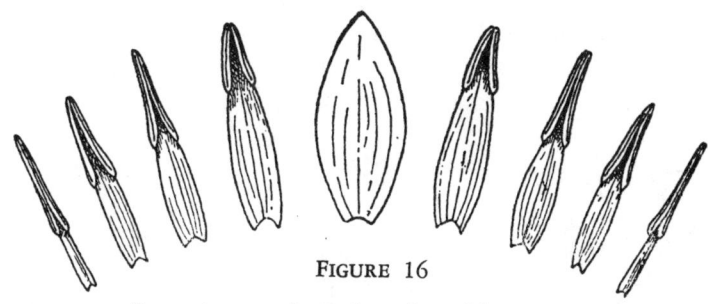

FIGURE 16

Successive transformation of petal into stamen,
in white water lily, *Nymphaea alba*.

genuine and but slightly modified petal contracts at the upper margin, allowing an anther, in which the rest of the petal takes the place of a filament, to come into view [Fig. 15].

48. In flowers that are oftentimes double, we can study this transition in all its stages. Within the completely formed and tinted petals of several rose varieties, there will appear other petals, contracted sometimes in the middle and sometimes at the side. The contraction [Fig. 16] is effected by a little callosity more or less like a perfect anther in appearance, and the leaf approaches the simpler form of stamen in proportion to the degree it was contracted. In some varieties of double poppies, completely developed anthers are supported by slightly modified petals of very full corollas; in other varieties, anther-like callosities draw the petals more or less together.

49. If all the stamens are transformed into petals, the flowers be-

come sterile; however, if staminal organs actually develop in a flower in the course of the doubling process, fertilization proceeds.

50. And so a staminal mechanism originates when the organs which we previously saw expanding as petals now make their appearance in a highly contracted and, at the same time, highly refined state. The interpretation we presented above is thereby corroborated, and we become increasingly alert to the alternating processes of contraction and expansion by means of which Nature ultimately reaches her goal.

VII. *Nectaries*

51. Abrupt as the transition from corolla to staminal organ is in some plants, we nevertheless observe that Nature cannot always traverse this distance in *one* stride. Rather, she produces intermediate organs, which sometimes approach one part in form and function, and sometimes another. Although these organs vary greatly in structure, they can be

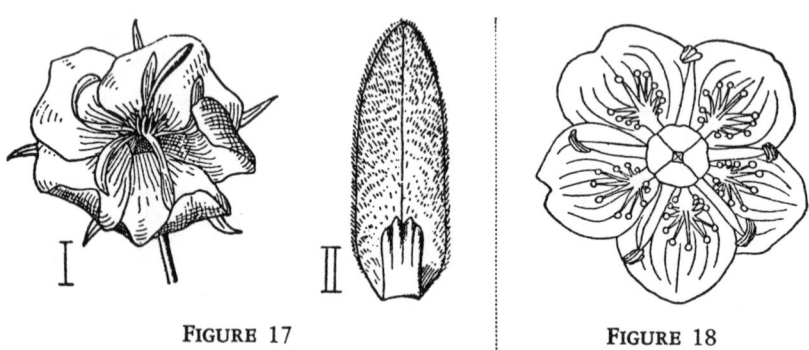

FIGURE 17   FIGURE 18

FIG. 17—I. Flower of *Pentapetes phoenica*. II. Petal of *Kiggellaria africana* with basal scale. FIG. 18—Flower of *Parnassia*, showing nectaries between stamens.

united for the most part under the concept that they are *gradual transitions from the sepals*[23] *to the stamens*.

52. Most of the diversely fashioned organs which Linné designated by the name *nectaries* may be grouped together according to this concept. And, once again, we have occasion to admire the keen mind of that extraordinary man, who, without definitely knowing the function

---

[23] Generally regarded as a slip of the pen for "petals," although Soret's translation, done under Goethe's supervision, has *feuilles du calice*. See p. 254.

ON MORPHOLOGY 49

of these apparently quite diverse organs, relied upon intuition and ventured to classify them under one name.

53. Moreover, various petals show their kinship with the stamens in being provided, without marked change in form, with little depressions or glands which secrete a honey-like juice. That this juice is a fertilizing fluid, as yet imperfect and not fully determinate, we can more or less conjecture from the ideas already set forth here; and this conjecture will reach an even higher degree of probability for reasons which we shall bring forward later.

54. Now the so-called nectaries may also make their appearance as independent parts, in which case they sometimes approximate the petals

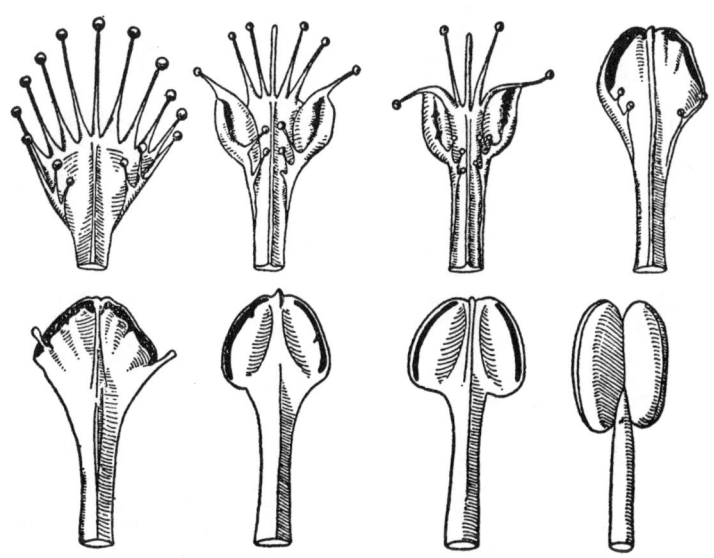

FIGURE 19

Intermediate forms of stamens and nectaries
in *Parnassia* (after Wettstein).

in structure and sometimes the stamens. In the nectaries of *Parnassia*, for example, the thirteen rays and like number of red globules greatly resemble the stamens [Figs. 18, 19]. Other nectaries appear as filaments without anthers, as in *Vallisneria* and *Fevillea*; in *Pentapetes* we find them in a circle, alternating regularly with the stamens and, indeed, already in leaf form [Fig. 17, I]. Also, in systematic classification they are listed as petal-like sterile filaments, *filamenta castrata petaliformia*. Similar fluctuating forms are seen too in *Kiggellaria* [Fig. 17, II] and in the passion flower.

55. Likewise, the coronas themselves appear to us to deserve the name nectaries in the sense explained above. For, although the forma-

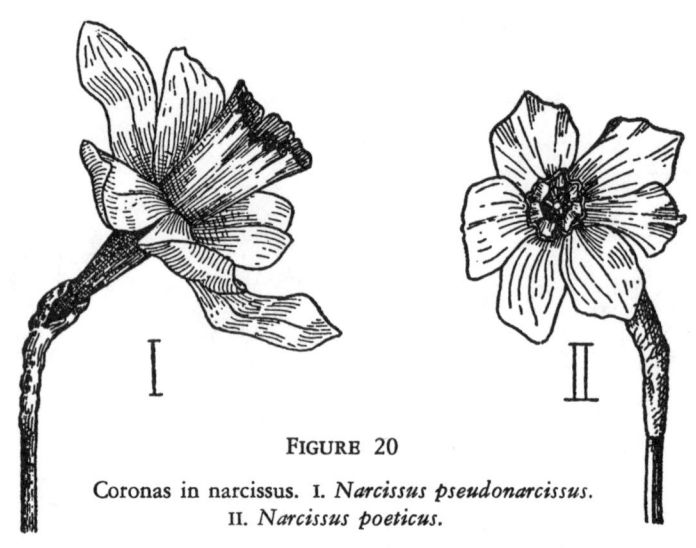

FIGURE 20

Coronas in narcissus. I. *Narcissus pseudonarcissus.*
II. *Narcissus poeticus.*

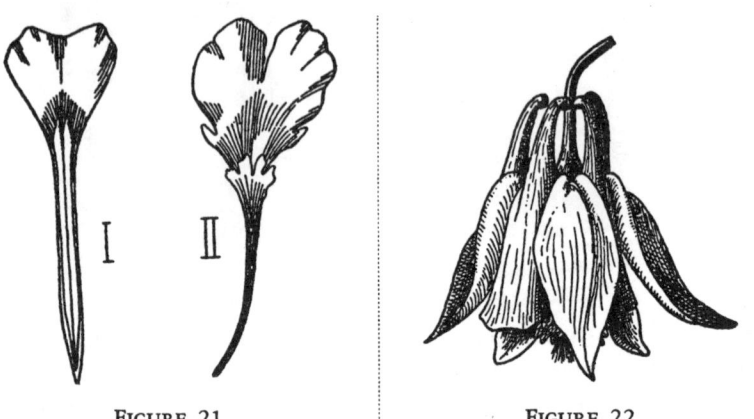

FIGURE 21   FIGURE 22

FIG. 21—Coronas on petals. I. *Agrostemma Coronaria.* II. *Melandrium rubrum.* FIG. 22—*Aquilegia,* petals alternating with spurred nectaries.

ON MORPHOLOGY

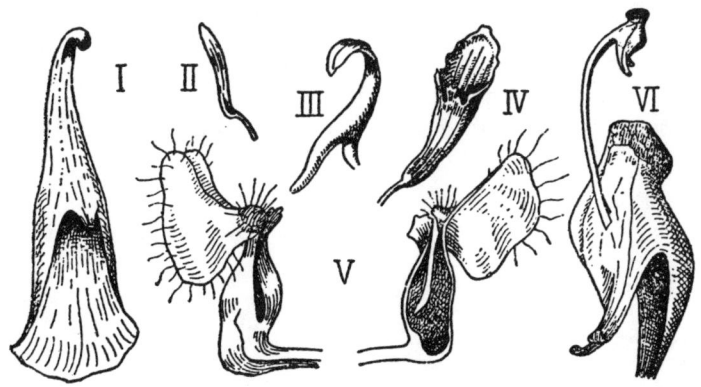

FIGURE 23

Nectaries of crowfoot plants, showing their leaf character. I. *Aquilegia* with spur. II. *Trollius europaeus*, with little indentation instead of spur. III. *Delphinium Staphisagria* with spur. IV. *Helleborus niger*, cornucopiate. V. *Nigella damascena*, in entirety and in lengthwise section, with hollow depression and cover above it. VI. *Aconitum Napellus*, nectary long-stemmed, visor-shaped, with sac-like hollow.

tion of the petals takes place through an expansion, the coronas are formed by a contraction; in other words, in the same way as the stamens. Thus, within the complete, expanded corollas we see smaller contracted coronas, for example, in *Narcissus* [Fig. 20], *Nerium*, and *Agrostemma* [Fig. 21].

56. In various other genera we see even more striking and remarkable transformations of petals. We notice that the petals of various flowers have a little indentation at the base of the inner surface, which is filled with honey-like juice. This little hollow receptacle, as it becomes deeper in other genera and species, produces a spurlike or hornlike elongation on the back surface, and the form of the rest of the petal is immediately more or less modified. We can note this clearly in various species and varieties of columbine [Figs. 22, 23].

57. For example, this organ is found in the highest degree of modification in *Aconitum* and *Nigella*, yet even here its similarity to the leaf can still be recognized with a little study [Figs. 23, 24, 25]. In *Nigella*, especially, the nectaries readily grow out again into petals, the flower becoming double through their transformation. In *Aconitum*, some care in examination reveals the similarity of the nectaries to the arched petals below which they are hidden.

58. Having just mentioned that nectaries are approximations of petals and stamens, we take the occasion to make some remarks about irregular flowers. For example, while the five outer leaves of *Melianthus* [Fig. 26] might be listed as true petals, the five inner ones might be described as a corona consisting of six nectaries, among which the nearest

FIGURE 24

Sketches by Goethe of various flower parts, among them nectaries of *Aconitum* and *Delphinium*.

approach to petal form is represented by the uppermost one, and the greatest divergence by the lowest, which is indeed already called a nectary. In the same sense, the keel of papilionaceous flowers might be called a nectary, since among the petals of these flowers it is the keel which most nearly approaches the stamen in form and which most widely diverges from the leaf shape of the so-called standard [Fig. 27]. In this way we can easily explain the brushlike bodies attached to the end of the keel in several varieties of *Polygala*, and thus gain a clear understanding for the classification of these parts [Fig. 28].

Tulip, illustrating case of leaf which is both stem leaf and petal. Original done in color under Goethe's supervision.

PLATE III

Perfoliate rose. Original done in color under Goethe's supervision.

PLATE IV

ON MORPHOLOGY

59. It should hardly be necessary here to point out that these remarks are not intended to confuse what has already been laboriously sorted out and classified by observers and systematists. They are merely intended to make anomalous plant formations more understandable.

VIII. *Additional Notes on the Staminal Organs*

60. Microscopic observations have shown beyond all doubt that the reproductive organs of the plant, like the other parts, are produced by means of the spiral vessels. From this fact we derive an argument for

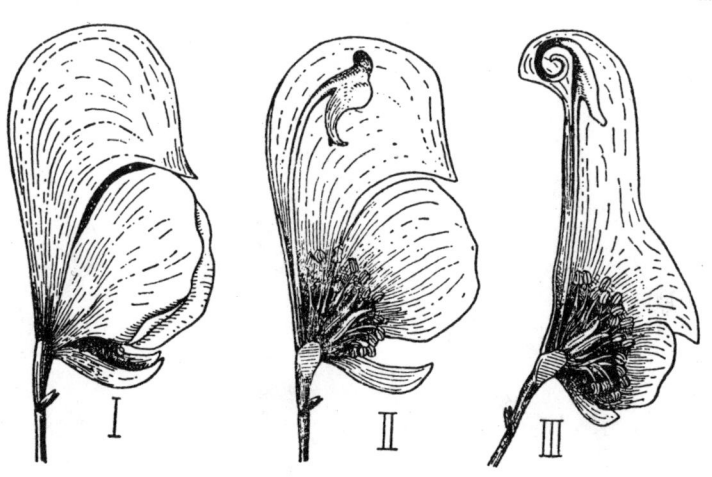

FIGURE 25

I. Flower of *Aconitum Napellus*. II. The same with nectary visible. III. Flower of *Aconitum lycoctonum*, lengthwise section, with spiral spur.

the essential identity of various plant parts that have hitherto appeared to us in such diverse forms.

61. Inasmuch as the spiral vessels are located in the center of the sap-vessel bundles, and are enclosed by them, we can picture that strong contraction somewhat more clearly if we imagine the spiral vessels, which actually appear to us like elastic springs, at the height of their power, when they are dominant and the expansion of the sap vessels subordinate.

62. The shortened vascular bundles can no longer expand; they no longer seek each other out and no longer build a network through anastomosis. The vesicles that usually fill up the interstices of the net can

no longer be formed; all factors contributing to the horizontal expansion of the leaves, sepals, and petals disappear completely; and a weak, extremely simple filament arises.

63. Indeed, the fragile little membranes of the anthers, between which the extremely delicate vessels now terminate, are just barely able to take form. If, then, we assume that those vessels which formerly lengthened, expanded, and again sought each other out, are at present in a highly contracted state; if we see the highly developed pollen emerging from them and replacing through its activity what the vessels that produce it have lost in ability to expand; if the liberated pollen now seeks out the female parts, which by a similar operation of Nature have been advancing toward the stamens; if the pollen firmly attaches itself to these female parts and transmits its influences to them: then we are not dis-

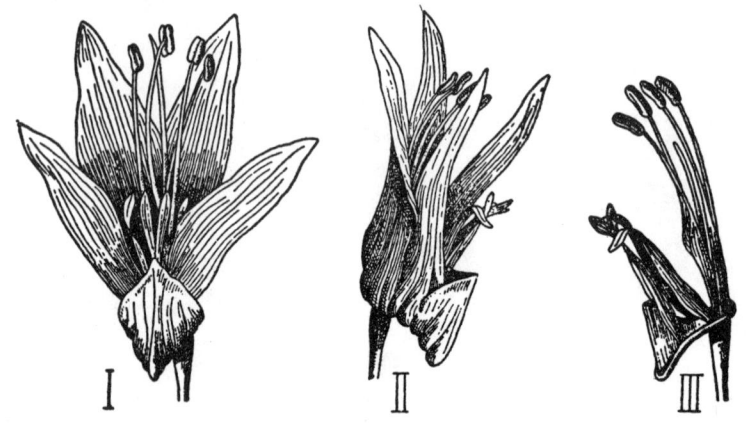

FIGURE 26

Flower of *Melianthus major* L. I. Front view. II. Side view. III. Side view, calyx removed, the slipper-like nectary visible.

inclined to call the union of the two sexes an idealized anastomosis, and we believe we have brought closer together, for a moment at least, the concepts of growth and generation.

64. The fine substance developing in the anthers appears to us like a powder; however, these little pollen grains are merely vessels in which extremely refined sap is stored. We therefore agree with those who maintain that this sap is absorbed by the pistils to which the pollen grains adhere, and that fertilization is brought about in this way. This is all the

more probable inasmuch as some plants form no pollen, but merely a fluid instead.

65. We recall here the honey-like secretion of the nectaries and its apparent connection with the more refined fluid of the seed vesicles. Perhaps the nectaries are preparatory organs; perhaps their honey-like fluid is absorbed by the stamens, becoming more fixed and developed. This

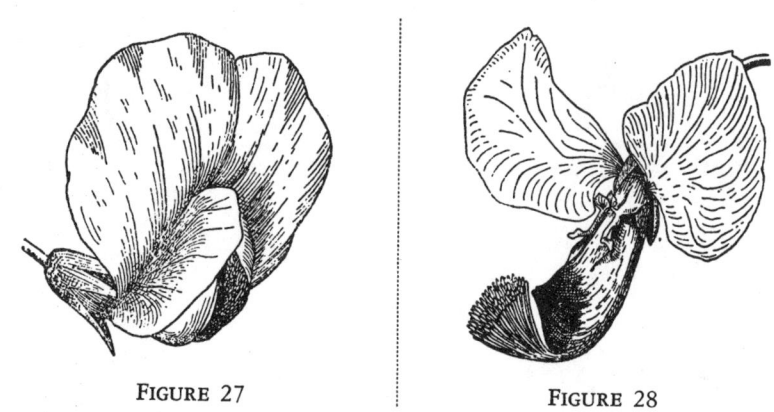

FIGURE 27   FIGURE 28

FIG. 27—Flower of pea, illustrating papilionaceous flowers; keel between wings, latter covered by standard. FIG. 28—Flower of *Polygala myrtifolia*.

view is quite plausible in that the secretion is no longer to be seen after fertilization.

66. We shall give at least cursory mention here to the fact that filaments as well as anthers are fused in various ways, providing us with the most wonderful examples of the anastomosis we have referred to several times and of the connection of plant parts which at their origin were distinctly separate.

IX. *Formation of the Style*

67. Up to this point I have endeavored to show, insofar as it was possible, the inner identity of various plant parts that develop successively and with greatest diversity in outward form. That it is now my intention to explain the structure of the female parts in this same manner, may easily be surmised.

68. First we shall observe the style separated from the fruit, as

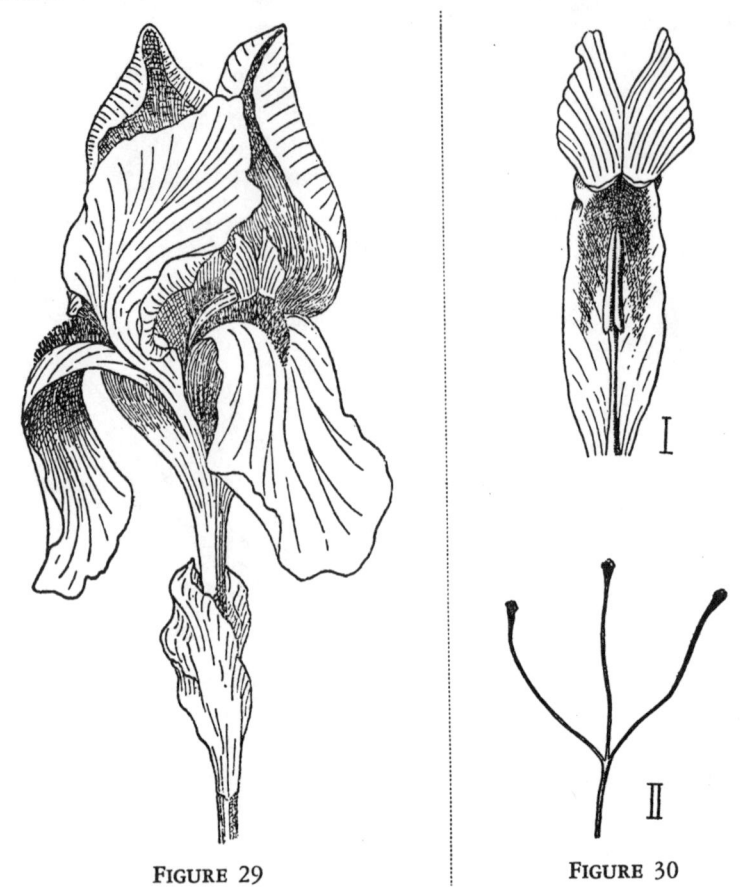

FIGURE 29   FIGURE 30

FIG. 29—*Iris germanica,* showing petal-like stigmas above the down-turned petals. FIG. 30—I. Iris pistil with stamen. II. Style and stigma of *Crocus sativus L.*

---

indeed we often find it in Nature; and we can do this very readily, for in that form the style displays its distinctive character.

69. For instance, we see the style in the very same stage of growth in which we found the stamens. These latter, we were able to observe, arose through contraction. Now we see that this is frequently true of styles also, and we find them, if not always the same size as the stamens, nevertheless only slightly longer or shorter. In many cases the style almost resembles a filament without an anther, and the relationship of

ON MORPHOLOGY 59

their external forms is closer than that of the other parts. Since both styles and filaments are produced by spiral vessels, we see all the more clearly that the female part has as little claim to being a special organ as has the male part; and when we are able by means of this viewpoint to visualize the exact relationship of male and female parts, we find the idea of applying the term anastomosis to fertilization more appropriate and more illuminating.

70. Very often we find the style produced through the fusion of several individual styles, and the composite parts are sometimes scarcely recognizable, for not even at the tips are they always distinct from one another. This fusing, the operation of which we have frequently noticed, is most likely to occur in the style; indeed, it *must* occur here, for the delicate parts are pressed together in the center of the flower before their development is complete, and they can thus associate most intimately.

71. Nature shows us more or less clearly in various normal cases the close relationship of the style to the preceding parts of the flower.

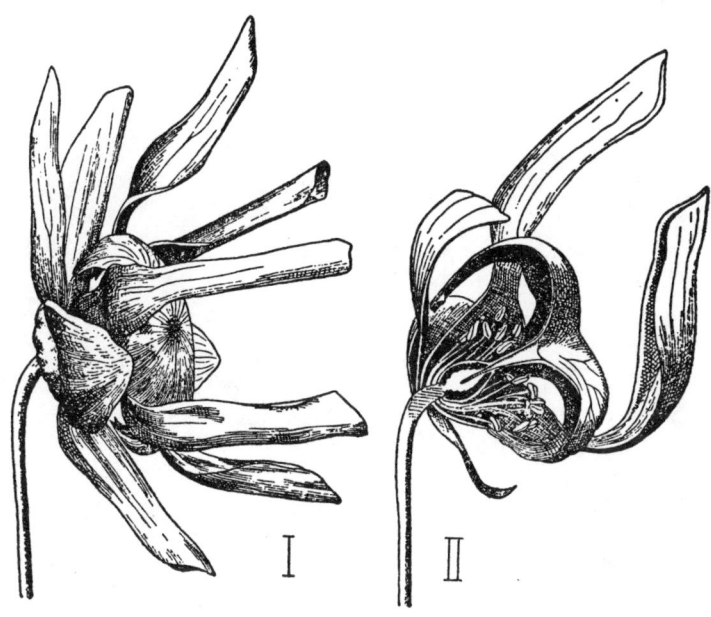

FIGURE 31

Sarracenia (after Wettstein). I. Flower in entirety, with umbrella-shaped stigma. II. Flower in lengthwise section.

For example, style and stigma of the iris present to the eye the complete form of a petal [Figs. 29; 30, I]. The umbrella-shaped stigma of *Sarracenia* does not reveal so strikingly that it is put together from several leaves, though its color is still the green of the leaves [Fig. 31]. If we use a microscope, we see that several stigmas, those of the crocus [Fig 30, II] and *Zannichellia* for instance, have the form of perfect calyxes of one or several leaves.

72. Retrogressively, Nature often shows us cases in which the styles and stigmas are again transformed into petals; for example, *Ranunculus asiaticus* is doubled when the stigmas and pistils of the fruit receptacles are transformed into genuine petals, while the stamens directly below the corolla are often unchanged [Fig. 32]. Several other important cases will be cited below.

73. By repeating here a remark made earlier, that styles and stamens represent the same stage of development, we can further clarify the cause of this alternate expansion and contraction. From seed to fullest development of stem leaves we noted first an expansion; thereupon we saw the calyx developing through contraction, the petals through expansion, and

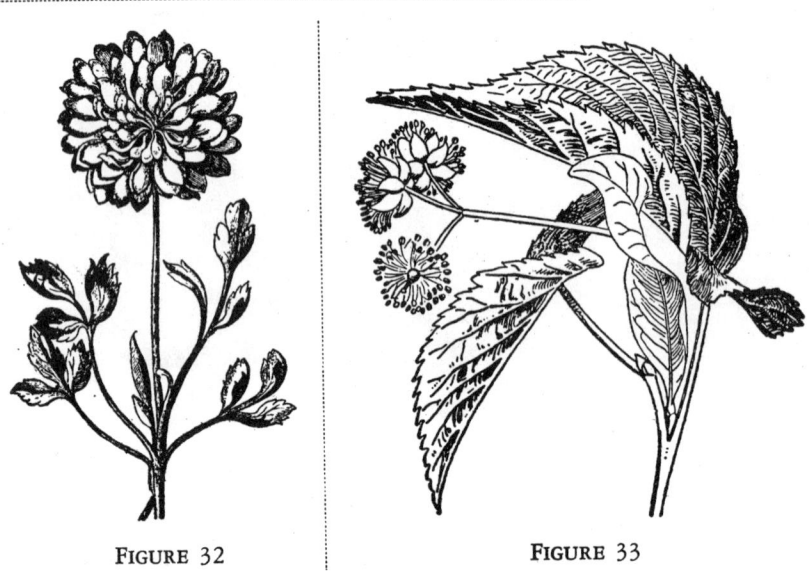

FIGURE 32   FIGURE 33

FIG. 32—*Ranunculus asiaticus,* with doubled flower. FIG. 33—Twig of linden, *Tilia platyphyllos,* showing little stalk emerging from midrib of leaf.

ON MORPHOLOGY 61

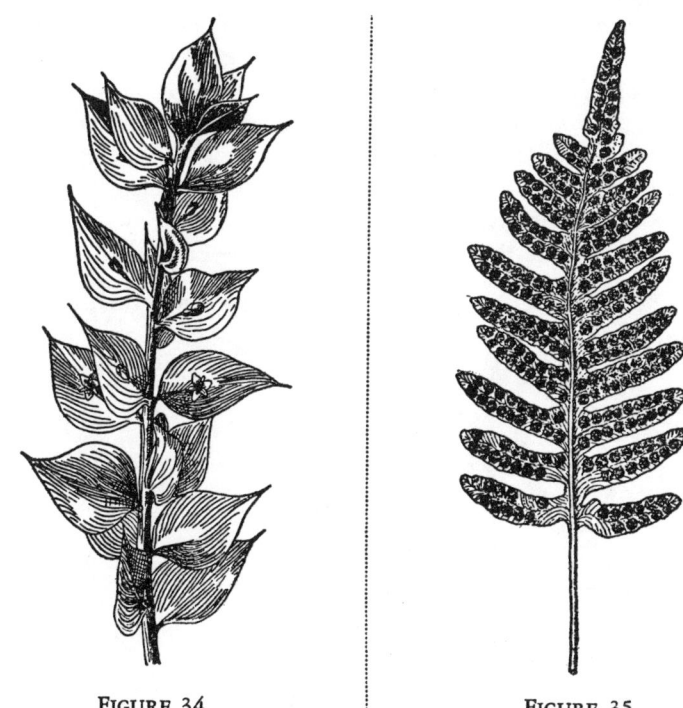

FIGURE 34   FIGURE 35

FIG. 34—Butcher's broom, *Ruscus aculeatus*. FIG. 35—Leaf of fern, *Polypodium vulgare*, lower side, thickly set with sori.

the sexual organs again through contraction; and soon we shall become aware of the maximum expansion in the fruit and the maximum concentration in the seed. In these six steps Nature ceaselessly carries on her eternal work of reproducing the plants by means of two sexes.

x. *The Fruits*

74. We shall now have to examine the fruits, and we shall soon be convinced that they have the same origin and are subject to the same laws. We are speaking specifically of the capsules which Nature forms to enclose the so-called covered seeds, or rather to develop from their interior, through fertilization, an indeterminate number of seeds. That these receptacles can likewise be explained on the basis of the nature and organization of the parts previously reviewed, can be readily demonstrated.

75. Again it is retrogressive metamorphosis that attracts our attention to this law of Nature. For example, in pinks—flowers whose very degeneracy has brought them fame and popularity—we often notice that the seed capsules are again transformed into calyx-like leaves and that the styles at the top decrease proportionately in length. Indeed, there are pinks in which the fruit shell is transformed into a genuine and perfected calyx. In these instances, the tips of the calyx-notches still bear delicate traces of the styles and stigmas, and in place of seeds a more or less perfect corolla develops from the interior of the second calyx.

76. Furthermore, even in regular and unchanging formations, Nature reveals in a diversity of ways the fertility that is latent in a leaf. Thus a linden leaf—modified, to be sure, yet completely recognizable—will

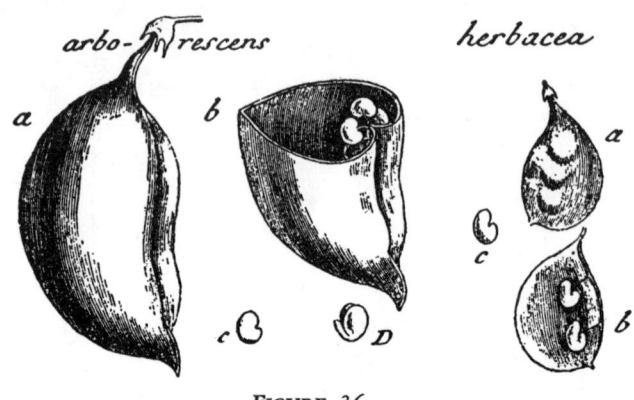

FIGURE 36

Hulls of *Colutea arborescens* and *Colutea herbacea*, showing their leaf character (after Gaertner).

produce from its midrib a little stalk and on it a perfect flower and fruit [Fig. 33]. In *Ruscus,* the manner in which the flowers and fruit rest upon the leaves is still more striking [Fig. 34].

77. Even more forcefully, almost abnormally, the direct fertility of stem leaves is put before our eyes in the case of ferns. From some inner force, and perhaps without direct interaction of two sexes, ferns develop and disseminate countless seeds, that is to say, germlings capable of growth [Fig. 35]. In this case, a leaf vies in fertility with a spreading plant or with a large and luxuriantly branching tree.

78. If we keep these observations in mind, we shall not fail to

ON MORPHOLOGY

recognize the leaf character of the seed containers, despite their diverse forms, their special modification, and their correlations. For example, the hull would prove to be a simple, folded leaf deformed at the margin [Fig. 36], the pods to consist of several leaves fused one above the other,

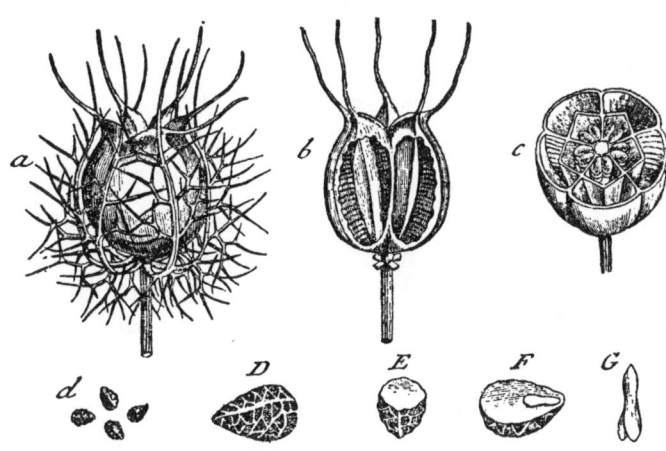

FIGURE 37

Fruit capsules of *Nigella damascena* (after Gaertner).

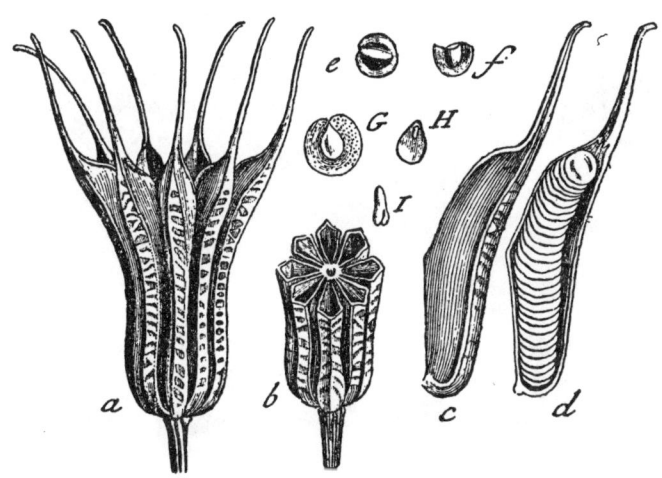

FIGURE 38

Nigella orientalis (after Gaertner).

and the composite cases to consist of several leaves grouped around a central point, their inner parts facing each other and their margins uniting [Figs. 39, 40]. We may have visual proof of this by observing such composite capsules when they spring apart after maturity, for then each part looks like an opened hull or pod. Likewise, we see a similar operation occurring regularly in various species of one and the same genus; for example, the fruit capsules of *Nigella orientalis* are gathered around an axis in the form of half-fused hulls [Fig. 38], whereas in *Nigella damascena* they are completely fused [Fig. 37].

79. Nature obscures this similarity to the leaf most when she makes the seed containers soft and juicy or firm and woody; however, the similarity will not escape our attention if we contrive to follow it in all its transitional stages. Here it is sufficient to have outlined the general conception involved, and to point out corresponding examples in Nature. Later, however, this great diversity of seed capsules will again provide us with material for consideration.

80. The relationship of the seed capsules to the preceding parts is revealed also in the stigma, which, in many instances, rests directly on the capsule and is inextricably bound up with it. We have previously indicated the relationship of the stigma to the leaf form and may properly bring it forward again as an example here, for in double poppies it can

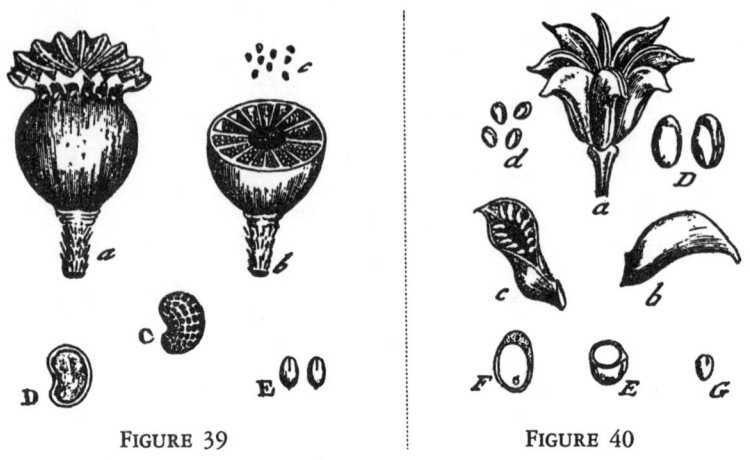

FIGURE 39   FIGURE 40

FIG. 39—Fruit capsule of poppy, *Papaver orientale* (after Gaertner).
FIG. 40—Fruit of marsh marigold, *Caltha palustris* (after Gaertner).

be noted that the stigmas of the seed capsules are transformed into delicate, colored leaflets completely resembling petals.

81. The last and greatest expansion which the plant undertakes during its growth, occurs in the fruit. Both in inner strength and outer form, this expansion is often very great, indeed enormous. It usually takes place after fertilization and thus it seems as if the seed, which by now has assumed a definite trend, directs the saps necessary for its growth

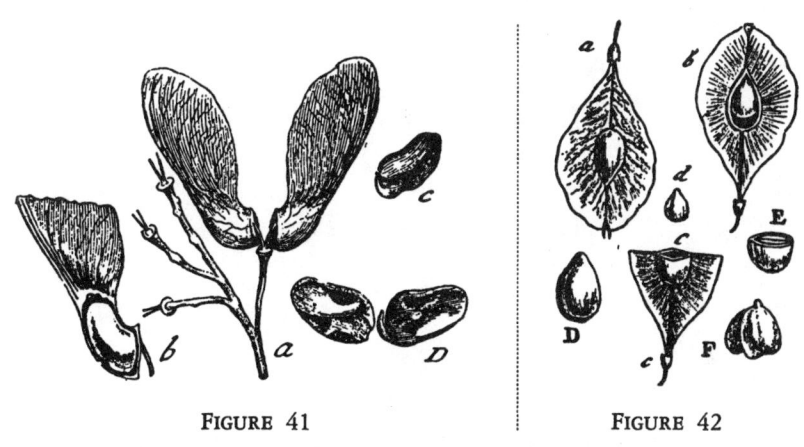

FIGURE 41   FIGURE 42

FIG. 41—Fruit of maple, *Acer tataricum* (after Gaertner).
FIG. 42—Fruit of elm, *Ulmus americana* (after Gaertner).

chiefly to the seed capsules, by attracting saps from the plant as a whole. In this way the vessels are nourished and dilated, often distending and expanding to an extremely high degree. From the foregoing we may conclude that purer forms of gas play a great role in this process, and our deduction is verified empirically, by the fact that the swollen hulls of *Colutea* contain pure gas [Fig. 36].

XI. *The Proximate Hulls of the Seed*

82. Conversely, we find that the most extreme degree of contraction and interior elaboration occurs in the seed. It can be observed in various instances that the seed transforms leaves to create its proximate hulls, and that it more or less adjusts them to fit. Indeed, by virtue of its vigor it usually annexes them completely and totally changes their form. Since we have already seen several seeds developing from and

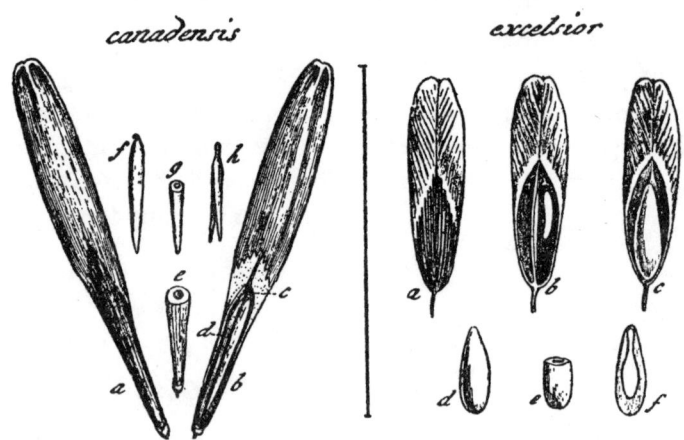

FIGURE 43

Fruit of ash, *Fraxinus* (after Gaertner).

within *one* leaf, we shall not be surprised to see an individual seed embryo garbing itself in a leafy cloak.

83. In many winged seeds we find vestiges of leaf forms which are imperfectly fitted to the seed, for example, in maple, elm, ash, and birch trees [Figs. 41, 42, 43, 44]. And the three differing circlets of the diversiform marigold seeds furnish a most remarkable example of how a seed embryo gradually contracts a broader hull and adjusts it to fit [Fig. 45]. The outermost circlet still retains a form akin to the sepals, except that a curvature of the sepal is caused by a seed rudiment that expands the ribs, and the resultant concavity is divided lengthwise into two parts by a membrane. The next circlet is already altered somewhat; the wing and the membrane of the leaf have disappeared completely, while on the other hand the shape has lengthened to a lesser degree; the seed rudiment

FIGURE 44

Fruit of birch, *Betula nigra* (after Gaertner).

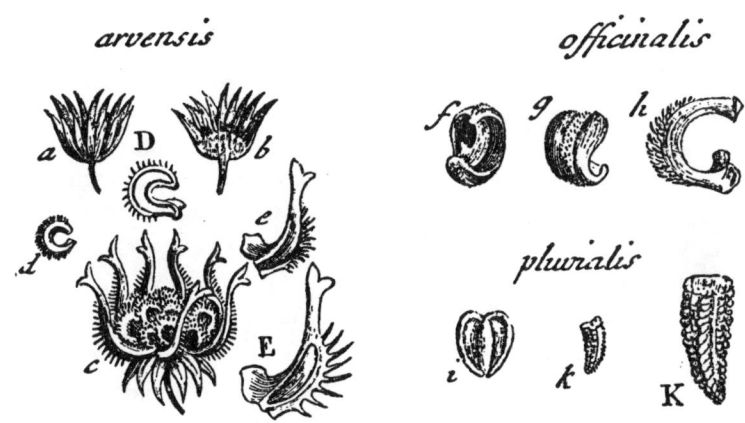

FIGURE 45

Fruit of marigold, *Calendula* (after Gaertner).

at the back emerges more clearly, and its little protuberances are in stronger evidence. These two rows of seeds do not appear to be fertile at all, or at most imperfectly so. They are succeeded by a third row, in characteristic strongly curved form, with a coat that is perfectly fitted and has developed its full complement of ridges and protuberances. Once again we see a vigorous contraction of expanded leaflike parts, effected this time by the internal force of the seed, just as we previously saw the petal contracted through the force of the anther.

XII. *Recapitulation and Transition*

84. We have attempted to follow as circumspectly as possible in the steps of Nature, accompanying the plant through all its outward transformations, from its development out of the seed to re-formation of the seed. Without claiming to be revealing the origin and mainsprings of Nature's processes, we have directed our attention to those manifestations of the forces by which the plant gradually transforms one and the same organ. To avoid losing the thread of our discourse we have considered the plant as an annual throughout; we have observed only the transformation of the leaves associated with the nodes and have derived all forms from them. However, to give this essay the necessary completeness, we must now also speak of the buds which lie hidden at the base of each leaf, and which apparently develop under certain circumstances and disappear completely under others.

XIII. Buds and Their Development

85. Each node receives from Nature the power to produce one or several buds. Moreover, this production occurs in the vicinity of the accompanying leaves, which seem to make provision for the formation and growth of the buds, and to participate in these processes [Fig. 46].

86. Upon the successive development of one node from another, and the formation of a leaf at each node with a bud in its vicinity, depends the first, simple, gradually progressing reproduction of vegetable life.

87. It is well known that such a bud greatly resembles the ripe seed in its activities and that often the whole form of the future plant may be discerned even more clearly in the bud than in the seed.

88. Even though the root tip is not easy to recognize in the bud, it is nevertheless present, just as it is in the seed; and it develops easily and quickly, especially under the influence of moisture.

89. The bud needs no cotyledons because it is connected with the mother plant, which is already completely organized, and receives sufficient nourishment from it as long as the two are linked. After separation, it continues to receive adequate nourishment from the new plant, if it has been grafted; or from the roots that form immediately, if it has been planted in the earth as a branch.

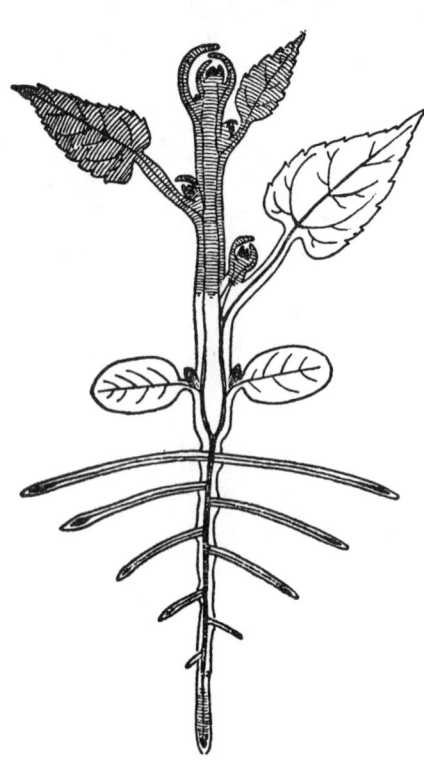

FIGURE 46

Diagramatic sketch of young dicotyledonous plant (after Sachs): white, parts already developed; hatched, those still in process of extension and growth; black, youngest parts. In axils of cotyledons and leaves proper, the "eyes," or buds, can be seen.

90. The bud consists of more or less developed nodes and leaves, whose function it is to foster the future growth of the plant. Thus the lateral branches, which develop from the nodes of plants, may be regarded as individual little plants which are set into the mother plant in the same way that the mother plant itself is fastened in the earth.

91. Buds and seeds have often been compared and contrasted, and recently this has been done with such unusual penetration and accuracy that we need merely direct the reader's attention to this work,* with unreserved approval.

92. We shall bring forward only the following from it. In complex plants, Nature clearly distinguishes between buds and seeds, but when we descend to the less complex ones, the difference between the buds and seeds tends to be lost even to the scrutiny of the sharpest observer. There are unmistakable seeds and unmistakable gemmae—genuinely fertilized seeds, which are isolated from the mother plant by the operation of two sexes, and gemmae, which merely emerge from the plant and detach themselves without visible cause—but the point where they coincide, though it can be grasped with the intellect, cannot possibly be seen with the eye.

93. Having fully pondered all this, we may venture to conclude that the seeds, although differentiated from the buds by being enclosed, and from the gemmae by the visible cause of their formation and detachment, are nevertheless closely related to both.

XIV. *Formation of Compound Flowers and Fruits*

94. We have previously sought to explain simple flowers, as well as seeds encased in capsules, by the transformation of nodal leaves; and on closer examination, it will be found that in such cases no buds develop, that the possibility of such a development is indeed completely cancelled. However, to explain both compound flowers and collective fruits arranged around a single cone, a single spindle, a single base, and so forth, we must refer to the development of the buds for assistance.

95. Very often we notice that stems do not long prepare or hold themselves in reserve for a single inflorescence, but force out their flowers at the nodes, frequently continuing to do so uninterruptedly to their tips.

---

*Gaertner, *De fructibus et seminibus plantarum,* ch. 1. [Joseph Gaertner, 1732–1791, whose fundamental work cited here by Goethe appeared in 1788–1791.]

Nevertheless, the accompanying phenomena may still be explained by the theory suggested above. All flowers which develop from buds are to be regarded as complete plants that rest upon the mother plant, just as the latter itself rests upon the earth. Even the first leaves of the branchlets, since they now receive purer saps from the nodes, appear much more developed than do the first leaves following upon the cotyledons in the mother plant. Indeed, formation of calyx and flower often becomes possible immediately.

96. With more abundant nourishment these same flowers that develop from buds would have become branches, thereby enduring the fate to which the mother stem would have to submit under similar circumstances.

97. When such flowers develop from node to node, we likewise observe in the leaves of the stem the same change we previously observed in the gradual transition to the calyx. These stem leaves contract more and more and finally disappear almost completely. They are then called bracts, since they deviate more or less from the leaf form. The stalk grows proportionately thinner, the nodes draw closer together, and all the phenomena mentioned previously again take place here, except that no sharply defined inflorescence occurs at the end of the stem, Nature having already exercised her right from bud to bud.

98. After carefully considering such a stem adorned with a flower at each node, we can easily explain a *collective inflorescence* if we keep in mind what was previously said about the origin of the calyx.

99. Nature forms a *common* calyx from *many* leaves, compressing and gathering them around a single axis. With the same strong vital impulse she develops an *extremely long stem, whose buds all appear simultaneously in flower form and are crowded together as closely as possible,* each floret fertilizing the seed container already prepared below it. During this abnormal concentration the nodal leaves do not invariably disappear [Figs. 47, 48]; in thistles, the leaflet faithfully follows upon the floret, which develops from the neighboring bud. We suggest that the form of *Dipsacus laciniatus* be examined in the light of this paragraph [Figs. 49, 50]. In many grasses each floret is accompanied by such a leaflet, which in this case is called a glume.

100. In this way it will be made clear to us that the seeds which develop around a *common inflorescence are true buds, formed and elaborated through the activity of two sexes.* If we lay firm hold on this

concept, examining and comparing the growth and fructification of several plants in the light of it, we shall have the evidence of our own eyes to convince us.

FIGURE 47 FIGURE 48

FIG. 47—Star-of-Bethlehem (*Ornithogalum minus*), representative of plants with "no definite inflorescence"; flowers in axils of bracts. From *Hortus Eystettensis.* FIG. 48—Orchis, representative of plants with a "collective inflorescence"; florets standing in the axils of the bracts. From *Hortus Eystettensis.*

101. Moreover, it will not be difficult then to explain the fructification of encased or unencased seeds in the center of a single flower, often gathered around an axis. For it is quite immaterial whether a single flower encircle a common fructification, the fused pistils absorbing the generative secretions from the anthers of the flower and directing them

into the seeds; or whether each seed be provided with its own pistil, its own anthers, and its own petals.

102. We are convinced that with some practice it would not be difficult to account for the diversified forms of flowers and fruits in this manner. To be sure, the conceptions established above—of expansion and contraction, compression and anastomosis—would have to be manipulated as expertly as algebraic formulae, and would have to be applied in the right places. A great deal would depend upon accurate observation

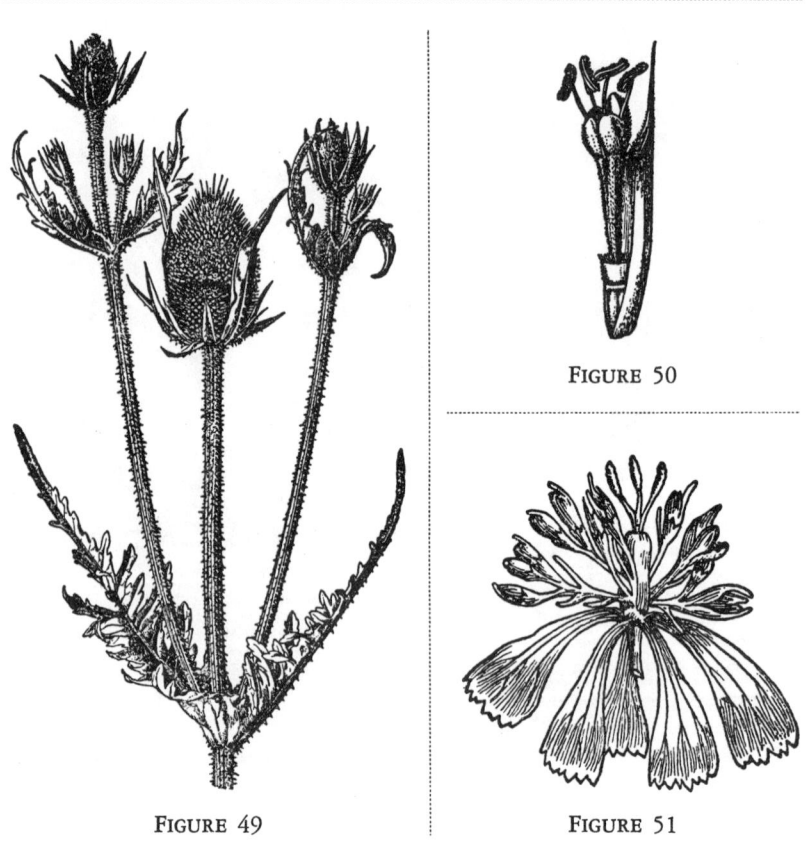

FIGURE 49

FIGURE 50

FIGURE 51

FIG. 49—*Dipsacus laciniatus*. Apical section of flowering plant with numerous florets crowded into the head-shaped inflorescence (after Hegi). FIG. 50—Floret of *Dipsacus laciniatus*, with bract, a so-called palea (after Hegi). FIG. 51—Perfoliate pink, calyx removed, petals turned back to show new flowers developing from petal area.

ON MORPHOLOGY 73

and comparison of the various steps which Nature takes in the formation of genera, species, and varieties, as well as in the growth of each individual plant. Hence, a collection of illustrations, arranged with this end in mind, and the application of botanical terminology to various plant parts would be desirable from this point of view alone, and not without use. Two cases of perfoliate flowers, which merely seem to bear out the theory cited above, would furnish conclusive proof when presented graphically to the eye.

XV. *The Perfoliate Rose*

103. Everything that we have hitherto sought to understand through imagination and intellect alone, we find most clearly exemplified in the perfoliate rose [Plate IV]. Calyx and corolla are arranged and developed around the axis; however, the seed container in the center is not contracted, nor are the male and female sexual parts *arranged* on and around it in ordered sequence; instead, the stalk shoots *upward* again, *half red* and *half green;* smaller, dark-red, folded petals, some bearing traces of anthers, develop successively on it. The stalk keeps on growing; thorns even appear on it again; the individual colored petals that follow become smaller, and finally are transformed before our eyes into half-red, half-green stem leaves; a succession of regular nodes is formed, from the buds of which there again emerge little rosebuds—albeit imperfect ones.

104. This same example also gives us another visible proof of a point made earlier, namely, that all calyxes are merely floral leaves whose margins are contracted. For here the calyx, regular and collected about the axis, consists of five fully developed compound leaves of three and five leaflets, of the kind that rose branches usually produce at their nodes.

XVI. *The Perfoliate Pink*

105. If we have observed this phenomenon correctly, another that appears in a perfoliate pink will seem to us almost more remarkable [Fig. 51]. We see a complete flower, provided with a calyx and a double corolla besides, terminating in the center with a seed capsule—which, however, is not completely developed. From the sides of the corolla four complete new flowers develop, separated from the mother flower by stems with three or more nodes. The new flowers have calyxes and are likewise doubled, not so much by single petals, however, as by corollas whose claws have coalesced, or more usually by petals which are

joined like branchlets and are assembled around a stalk. This abnormal development notwithstanding, the filaments and anthers are present in some. The fruit coverings and their styles are visible and the seed receptacles are again transformed into leaves. Indeed, in one of these flowers the seed cases were collected into a complete calyx, containing in its turn the rudiment of a completely doubled flower.

106. In the perfoliate rose, we saw the development of a presumably half-defined flower, the re-emergence of the stem from its center, and the production of new leaves on this stem. In this pink—despite a well-formed calyx, a complete corolla, and *pistils placed in the very center*—we see *buds developing from the area of the petals* and displaying actual branches and flowers. And so both examples show that Nature usually completes her growth in the flowers, and closes the account here, so to speak, thus cutting short the possibility of an endless progression by single steps, in order that she may more speedily reach her goal through the formation of seeds.

XVII. *Linné's Theory of Anticipation*

107. If I have stumbled here and there on a road described as perilous and terrifying by one of my predecessors*—who, moreover, traversed it at the side of a great teacher; if I have not leveled it off sufficiently, nor cleared it of all obstacles for my successors: nevertheless, I hope I have not undertaken my endeavor in vain.

108. Now is the proper time to consider the theory which Linné formulated in clarification of these very phenomena. The same observations which evoked the present essay did not escape his sharp eye. If it is possible for us now to advance beyond the point where he left off, we owe it to the common efforts of numerous observers and thinkers who have cleared away many an obstacle from the path and who have dispersed many a prejudice. A detailed comparison of Linné's theory with the one set forth above would detain us too long. The initiated can easily make the comparison themselves; to make it understandable to those who have never studied the subject would involve too much detail. So we shall merely indicate in brief what it was that hindered Linné from striding forward to his goal.

109. He made his observations first on trees, those complex and long-lived plants. He noticed that a tree, placed in a rather large pot

---

* Ferber, *In Praefatione Dissertationis secundae "de Prolepsi Plantarum."*
[Johann Jakob Ferber, 1743–1790, professor of physics in St. Petersburg.]

and supplied excessively with nourishment, produced branch upon branch several years in succession, whereas the same tree, confined in a smaller pot, quickly bore fruits and flowers. He saw that in the latter case the usual successive development was sudden and concentrated. He therefore called this operation of Nature *prolepsis,* anticipation, because by means of the six steps outlined above, the plant appeared to advance by six years. And so he also developed his theory with respect to the buds of trees without giving much consideration to annual plants, for he evidently noticed that his theory did not fit the annuals so well as it did the trees. For according to his theory, one would have to assume that each annual plant, though actually destined by Nature to grow for six years, on reaching the flower and fruit formation, prematurely completes this prescribed period and thereupon dies.

110. Unlike Linné, we have followed first the growth of annuals, and can now easily make the application to perennials, for a bud unfolding on the oldest tree is to be regarded as an annual plant, even though it has developed from a stem which has long been in existence and may itself be of longer duration than a year.

111. The second circumstance that hindered Linné from going forward was that he interpreted too narrowly the various concentric zones of the plant body—the outer and inner bark, the wood, and the pith—as equally functioning parts, equally alive and essential; and that he attributed the origin of flowers and fruit parts to these various zones of the stem, because, like the stem zones, they appear to be enclosed by one another and to develop from one another. This, however, was only a superficial observation, one which is not confirmed on closer study. For example, the outer bark is unsuited to further production, and in perennial trees becomes a solidified and isolated mass towards the outside, while the wood becomes hard in the center. In many trees this outer bark is shed, and from others it may be removed without the slightest damage to the tree; it cannot, therefore, produce a calyx or any other living plant part. Actually, it is the second bark that contains all power of life and growth. To the extent that this is injured, growth is also disturbed; it is the second bark, we find on closer study, that produces all the exterior plant parts, gradually in the stem, or all at once in flowers and fruit. Yet, to this second bark Linné ascribed only the subordinate function of producing petals. To the wood, on the other hand, he ascribed the important task of producing the male pollen organs; and yet we can easily recognize the wood as a part brought to rest through solidification, and, though long-lasting, devoid of vitality. In conclusion, he believed

that the pith performed the prime function of producing the female sexual parts and numerous progeny. The doubts which others have raised regarding this supposed importance of the pith, and the reasons they have set forth, have weight and cogency likewise for me. It is only on the surface that styles and fruit seem to develop from the pith, for when we see these forms for the first time, they are in a soft, indeterminate, pithlike, parenchymatous state, and are crowded together in the exact center of the stem, where we are accustomed to see only pith.

XVIII. *Recapitulation*

112. I hope that the present attempt to explain the metamorphosis of plants may contribute something to the solution of these problems and provide occasion for additional comments and opinions. The observations on which my essay is based have already appeared singly and they have also been collected and classified.* It will soon be decided whether the step we have just taken is an approach toward the truth. As briefly as possible we shall summarize the chief results of the discussion up to this point.

113. When we consider a plant in relation to its vital force, we see this vitality manifesting itself in two ways: first, through *vegetative growth,* by development of stems and leaves; and next, through *reproduction,* which is completed in the formation of flower and fruit. If we examine the growth phase more closely, we see that the plant, as it vegetates and progresses from node to node, from leaf to leaf, is likewise carrying on a type of reproduction, which differs from that occurring in fruit and flower in that it is *successive* instead of sudden, appearing in a series of individual developments. Yet this vegetative force which exerts itself gradually is very closely related to the force which brings about a marked propagation in one step. Under certain circumstances a plant can be forced to *vegetate* continuously; and on the other hand, its flowering can be *accelerated.* The former situation occurs when there is a considerable influx of cruder saps, and the latter when more rarefied forces are preponderant.

114. By referring to vegetative growth as a successive reproduction, and to the formation of flowers and fruits as a simultaneous one, we have actually characterized the manner of their development also. A plant which vegetates, is expanding more or less: it develops a stalk or

---

*Batsch, *Introduction to the Science and History of Plants,* Part I, ch. 19. [See pp. 96, 155, 183, 189, 196, 217 for further reference to Batsch.]

stem, the distances from node to node are usually considerable, and its leaves spread out from the stem on all sides. Conversely, a plant which flowers, is contracting all its parts: increments in length and breadth are arrested, and all its organs, developing in close propinquity, are in a highly concentrated state.

115. Whether then the plant vegetates, blossoms, or bears fruit, it nevertheless is always the same organs, with varying functions and with frequent changes in form, that fulfill the dictates of Nature. The same organ which expanded on the stem as a leaf and assumed a highly diverse form, will contract in the calyx, expand again in the petal, contract in the reproductive organs, and expand for the last time as fruit.

116. This process of Nature is at the same time bound up with another, with the assembling of *various organs around one central point* in fixed numbers and proportions—greatly exceeded and variously modified, however, in some flowers and under certain conditions.

117. Similarly, anastomosis is in operation during the formation of flowers and fruit, closely uniting the compact and extremely delicate parts of the fructification, throughout their existence or for only part of it.

118. Yet these phenomena of *approach, centralization,* and *anastomosis* are not peculiar to flowers and fructifications alone; indeed, we can observe something similar in the cotyledons, and other plant parts will furnish us with abundant material for similar reflections in the sequel.

119. We have ventured to trace back to the leaf form those fruits in which the seeds are firmly enclosed, just as we sought to show that the organs of the vegetating and flowering plant, though seemingly dissimilar, all originate from a single organ, namely, the leaf, which usually develops at each node.

120. It is self-evident that we ought to have a general term with which to designate this diversely metamorphosed organ and with which to compare all manifestations of its form. At present we must be content to train ourselves to bring these manifestations into relationship in opposing directions, backward and forward. For we might equally well say that a stamen is a contracted petal, as that a petal is a stamen in a state of expansion; or that a sepal is a contracted stem leaf approaching a certain stage of refinement, as that a stem leaf is a sepal expanded by the influx of cruder saps.

121. We may likewise say of the stem that it is an expanded

flowering and fruiting phase, just as we have predicated of the latter that it is a contracted stem.

122. Moreover, I have at the close of the treatise considered also the development of *buds* and have thereby attempted to explain compound flowers and unenclosed fruits as well.

123. In this way, then, I have endeavored to set forth, as clearly and completely as I could, a theory which to me has much that is convincing. However, if my theory has not been conclusively demonstrated, if it should still contain contradictions, or if the method of interpretation it employs should not seem at all times applicable: all the more shall I consider it my duty to take note of all criticism and to give the material more exact and extended treatment in the sequel, thereby making this approach to the subject more graphic, and winning for it more universal approbation than can perhaps be expected at present.

Metamorphosis of Plants—Second Essay[24]

Introduction

1. Far removed as forms of organized life are from one another, we nevertheless find that they have certain characteristics in common, and that certain parts may be compared with one another. Used correctly, this concept is the thread that will guide us through the labyrinth of living forms; misused, it would lead us astray and take us backward rather than forward on the path of science.

2. Since all living things are alike in that they have the power of reproducing their own kind, we are justified in searching for the organs of reproduction throughout the plant world, just as we do in animal life. Indeed we find them almost down to the lowest step of the plant realm, where they still continue to receive the attention of observers.

3. In addition to this most general characteristic, we find other characteristics closely connected with it which likewise permit comparison. Thus, generally speaking, the seed capsule may still be compared with the ovary, and the seed with the egg. However, if we carry this too far and attempt to compare the parts of a plant seed with those of a bird's egg or even animal offspring, it is my opinion that we shall stray as far

---

[24] Written about 1790; published posthumously in the great Weimar edition of Goethe's complete works, 1887–1912 (see "Bibliographical Note," p. 255).

from the truth as we were close to it in the beginning. To the great extent that plants differ from animals, the plant seed must differ from the egg or embryo.

4. For that reason, comparisons of cotyledons with the placenta, and of various seed hulls with the amnion of animal births, are deceptive and all the more dangerous because they deter us from more accurate knowledge of the nature and characteristics of such parts.

Nevertheless, it was inevitable that the comparison should be pressed too far, since Nature actually gives some justification for it. For instance, the term marrow was given, justifiably perhaps, to the network that fills up the hollow stems of some plants, and this network was compared with the marrow of animal bones. However, the wrong conclusion was drawn, namely, that marrow is an essential part of plant bodies.[25] Marrow was sought for, and found, in places where it did not exist; powers and influence were attributed to it that it did not possess, all because the concept of marrow in human bones was adhered to. All this attained greater importance than it merited through the poetically imaginative terminology that had made its way into science.

See the essay on animal forms.[26]

5. Things went even further. For the convenience of imagination and for the fostering of certain fanatical religious ideas, an attempt was made to trace all things back to one, and to find all in everything. Thus muscles, veins, lymph vessels, intestines, gullets, glands, and what not were found in plants.

See Agricola *Agriculture parfaite.*[27]

To be sure, these false observations were gradually displaced by more exact observations, especially microscopic ones. However, a great deal remains that ought to be cast away in the best interest of science.

6. It is undoubtedly proper here to recall other parables, since comparisons are made not only of one natural realm with another but even with objects from the universe as a whole. Great harm is done through this specious manipulating of the physiology of the three realms. For example, Linné calls flower petals "curtains of the nuptial bed," a parable that would do honor to a poet. But after all! The discovery of the true

---

[25] See par. 111, pp. 75-76.
[26] Goethe is referring to an osteological essay, unpublished during his lifetime.
[27] Georg Andreas Agricola, 1672-1738, *Agriculture parfaite* (Amsterdam, 1720). Agricola's theories on propagation and grafting were supported by an understructure of mysticism.

physiological nature of such parts is completely blocked in this way, just as it is by the convenient and false espousal of the theory of final causes.[28]

In my opinion, the chief concept underlying all observation of life— one from which we must not deviate—is that a creature is self-sufficient, that its parts are inevitably interrelated, and that nothing mechanical, as it were, is built up or produced from without, although it is true that the parts affect their environment and are in turn affected by it.

See the essay on animal forms.

7. This concept is the basis of my first essay on the metamorphosis of plants, a concept which I shall not lose sight of in this present exposition, nor in any other observation of living organisms which I may undertake. But I have already explained on another occasion that here it is not a question of whether the concept of final causes is convenient, or even indispensable, to some people, or whether it may not have good and useful results when applied to the moral realm. Rather, it is a question of whether it is an aid or a deterrent to physiologists in their study of organized bodies. I make bold to assert that it does deter them, therefore avoid it myself and consider it my duty to warn others against it. For, as Epictetus says, one should grasp an object by the smooth handle to make seizure easy. Moreover, the scientist may set his mind at rest and travel his path undisturbed, for the newer school of philosophy will not fail to consider it a duty to popularize this concept after the method outlined by their teacher,[29] at which time the scientist should not miss his opportunity to speak his mind on this score.

8. In the first essay, I endeavored to show that various plant parts develop from a wholly analogous organ,[30] which, although remaining basically the same, is modified and changed through progression.

9. Underlying this principle is another, namely, that a plant has the power of infinite reproduction through mere development of completely analogous parts. For example, when I lop off and plant a willow twig, later cutting off and replanting the next shoot, I can continue the process into infinity. Likewise, when I tear off and plant a stolon, it gives me new stolons without blossoming, and so forth ad infinitum.

10. The second empirical principle derived from this is that growth, which continues upward above the earth, cannot always go forward at

---

[28] For another instance of Goethe's dislike of teleological principles, see the essay, "An Attempt to Evolve a General Comparative Theory," pp. 81–84.
[29] Here Goethe's manuscript gives an incomplete reference to Kant's *Critique of Judgment*.
[30] See par. 115, p. 77.

the same pace, but must gradually change its form and give the parts a different function. This is the regular progressive metamorphosis of plants, which interests man most because his attention is usually focussed upon the resulting fruits and flowers.

11. The purpose of this present and second essay is to continue the foregoing reflections, clarifying them through illustrative cases, rendering them more graphic by the use of copper engravings, and lending them greater authority by citation of authors. Likewise, whatever is pertinent from the science of botany as a whole should be introduced, and the way prepared for further progress.[31]

An Attempt to Evolve a General Comparative Theory[32]

WHEN A SCIENCE appears to be slowing down and, despite the efforts of many energetic individuals, comes to a dead stop, the fault is often to be found in a certain basic concept that treats the subject too conventionally. Or the fault may lie in a terminology which, once introduced, is unconditionally approved and adopted by the great majority, and which is discarded with reluctance even by independent thinkers, and only as individuals in isolated cases.

From this general remark I shall proceed at once to the subject under consideration, in order that I may be as clear as possible at the very outset and avoid straying from my objective. For several centuries down to the present, we have been retarded in our philosophic views of natural phenomena by the idea that living organisms are created and shaped to certain ends by a teleological life force. Certain individuals have, however, vigorously opposed this concept and have pointed out the obstacles it puts into our path.

It may very well be that the concept is harmless in itself, acceptable to certain temperaments, and actually indispensable to some modes of thought, and I myself find it neither possible nor desirable to oppose it as a whole. It is, if I may so express myself, a trivial manner of thought,

---

[31] Here the projected essay comes to an end.
[32] Preliminary notes for an essay found among Goethe's literary effects and published posthumously in the Weimar edition. Apparently dictated in the early 1790's and not revised by Goethe himself, it was written by one secretary and contains pencilled corrections by another. It was further corrected by the Weimar editors.

and, like all trivial things, remains trivial precisely because human nature finds it so thoroughly adequate and convenient.

Man is accustomed to value things to the extent that they are useful to him, and since he is disposed by temperament and situation to consider himself the crowning creation of Nature, why should he not believe that he represents also her final purpose? Why should he not grant his vanity this little fallacy? He has need of and can make use of certain objects, and thus draws the conclusion that they were created expressly for him. Why should he not daringly resolve the contradictions he encounters, rather than retreat from desires he is experiencing at the moment? Why should he not call a plant a weed, when from his point of view it really ought not to exist? He will much more readily attribute the existence of thistles hampering his work in the field to the curse of an enraged benevolent spirit, or to the malice of a sinister one, than simply regard them as children of universal Nature, cherished as much by her as the wheat he carefully cultivates and values so highly. Indeed, the most moderate individuals, in their own estimation philosophically resigned, cannot advance beyond the idea that everything must at least ultimately redound to the benefit of mankind, or indeed that some additional power of this or that natural organism may yet be discovered to render it useful to man, in the form of medicine or otherwise. And since man values highest, in himself and others, those processes which are intentional and purposeful, he will ascribe intentions and purposes to Nature also, for his concept of Nature cannot possibly transcend the concept he has formed of himself.

If he further believes that everything in existence exists for his own sake, exists only as a tool and auxiliary instrument for his own life, it inevitably seems to him that Nature has proceeded intentionally and purposefully in creating tools for him, in the same manner that he himself has done. Thus the hunter who procures a gun for killing game cannot extol sufficiently the maternal solicitude of Nature in having created the dog at the very beginning of things, to enable him to retrieve the game. And there are additional reasons why it is generally impossible for man to drop this teleological concept, but from the example of botany alone we can see how necessary it is that the scientist who desires to reflect further on universal matters should give up this idea. To the science of botany, the most colorful and complex flowers, the most delectable and beautiful fruits, are not more valuable—indeed, in a certain sense are not worth as much—as a despised weed in its natural state, or a dried and seemingly worthless seed pod.

The natural scientist will inevitably have to rise above this trivial teleological concept. If he cannot rid himself entirely of the idea, he must at least free himself as much as possible.

The problem just discussed, of general concern to scientists, has only general importance to us also. Another problem, stemming directly from the first, concerns us somewhat more closely. Man, in considering all things with reference to himself, is obliged to assume that external forms are determined from within, and this assumption is all the easier for him in that no single living thing is conceivable without complete organization. Internally, this complete organization is clearly defined; thus it must find external conditions that are just as clear and definite, for its external existence is possible only under certain conditions and in certain situations. For instance, we see moving about on the earth, in the water, and in the air, the most varied forms of animals; these elements, according to popular interpretation, have been furnished to these creatures expressly in order that they may produce their various movements and preserve their various existences. But does not the original life force, or the wisdom of a reasoning creator customarily attributed to it, gain greater stature when we accept even its power as limited, and grant that it creates just as well *from* the outside as well as *toward* the outside. To say that the fish exists for the water, seems to me to say less than that the fish exists *in* water and *by means of* water; for this latter statement expresses much more clearly what is only darkly suggested in the first, namely, that the existence of creatures called fish is possible only if there exists an element called water, and that these fish not only exist but also develop there. The same thing holds true for all other creatures. This would then be the first and most general observation on the development *from within toward without* and *from without toward within;* the ultimate form is likewise the inner nucleus, which is variously developed through determination by the outer element. An animal possesses external usefulness precisely because it has been shaped from without as well as from within, and—more important and quite natural—because the external element can more readily adapt the external form to its own purposes than it can *re*shape the internal form. We can best see this in a species of seal whose exterior has taken on a great deal of the fish character while its skeleton still represents the perfect quadruped.

We shall not be detracting from either the original life force or an all-wise and all-powerful creator, if we assume that both work mediately, the latter at the beginning of all things, the first continuously throughout. For is it not in keeping with this great force that it should produce the simple simply and the complex complexly? Do we detract from its power in maintaining that it could no more produce fish without water, or birds without air, and other creatures without the earth, than we ourselves can imagine such creatures without these prerequisite elements? Do we not gain a lovelier glimpse into the mysterious structure of form—more and more generally acknowledged to be constructed according to a single pattern—if, having explored and recognized the single pattern in greater and greater detail, we now ask and explore the questions: What effect does a general element in its various modifications have upon one and the same form? What countereffect does the determined and determining form have upon those elements? What is the effect on the shape of the firm, the softer, the innermost, and the outermost parts? And, as mentioned before, what results do the elements in all their modifications bring about through height and depth, geographical zones and regions?

How much preliminary work has already been done in this very direction! All that is necessary is to understand and apply it.

And how fitting that Nature must always employ the same means to produce and maintain her creations! Thus we shall stride ahead in these directions. Heretofore recognizing only the unorganized and indeterminate elements as vehicles of the unorganized, we shall henceforward rise to higher levels of thought and once more look upon the organized world as a composite of many elements. The whole plant realm will again appear to us like a vast sea, as necessary for the contingent existence of the insect as the actual seas and rivers are to the contingent existence of fish. We shall observe an enormous number of living creatures born and nourished in the plant sea. Indeed, we shall eventually regard the whole animal kingdom once more as an element in which one species is supported on and by means of another, if not actually originating one from the other. We shall give up the habit of considering conditions and relationships as ends and purposes, and shall progress exclusively through knowledge of the way creative Nature is molded *from* all directions and *toward* all directions. And experience will convince us, as the progress of science has always proved, that the most solid, widespread, and utilitarian achievements on behalf of mankind are exclusively the result of great and altruistic effort, one that does not demand its pay at the end of the week as day laborers do, but on the other hand is also not obligated to produce practical results at the end of a year, nor of a decade, nor of a century.

Preliminary Notes for a Physiology of Plants[33]

Concepts of Physiology

THE METAMORPHOSIS of plants is the basis of the physiology of plants. It shows us the laws by which plants are formed.

It draws our attention to a twofold law: (1) the law of inner nature, whereby the plant has been constituted; (2) the law of environment, whereby the plant has been modified.

Botanical science, on the one hand, acquaints us with the diverse formation of the plant and its parts, and at the same time searches for the principles of its formation.

We grant that attempts to classify the great mass of plants into a system deserve the highest degree of approval only when such attempts are necessary, when they separate the immutable parts from the more or less accidental and mutable ones, and in that way shed more light upon the extremely close relationship of the various families. Nevertheless, those efforts are also praiseworthy that seek to recognize the law whereby the formations are produced; and even though it appears that human nature is incapable of grasping clearly the infinite diversity of the organization or the law whereby it works, still it is a fine thing to summon all one's powers and to amplify this field in two ways, by experience and reflection both.

We have seen that plants are propagated in different ways, but these various methods are to be regarded as modifications of a single basic method. Propagation and prolification, occurring through development of one organ from another, have been our chief concern in the subject of metamorphosis. We have seen that such organs, though outwardly changed from similarity to the greatest dissimilarity, have a virtual inner identity.

We have seen that this sprouting prolification cannot continue into infinity with the most complex plants, that instead it leads step by step to a culmination and, as though having attained the opposite pole of its power, brings forth another kind of propagation through seeds.

Characterization and limitation of the field in which we are working.

Phenomenon of organic structure.

Phenomenon of the simplest structure which appears to be a mere aggregation of parts, but often explainable just as well through evolution or epigenesis.

---

[33] Written in the middle of the 1790's as an outline for the projected complete work on morphology; published posthumously in the Weimar edition.

Increase of this phenomenon and uniting of this structure into a zoological unit.

Form.

Necessity of considering all expository methods together, not to thoroughly explore a thing and its nature, but to give at least some description of the phenomenon, and to impart to others what has been perceived and seen.

Those bodies called organic have the characteristic of producing their like by themselves or from themselves.

This is part of the concept of an organic being and we can give no further account of it.

The new and the similar is at the beginning always a part of the same thing, and in this sense proceeds from it, thus supporting the idea of evolution.[34] However, the new cannot develop from the old unless the old has reached perfection of a sort through a certain absorption of outer nourishment, thus supporting the idea of epigenesis. Both concepts are crude and coarse compared with the delicacy of the unfathomable phenomenon itself.

In an organic being, first the form as a whole strikes us, then its parts and their shape and combination.

Form in general and the relation and combination of the parts, insofar as they are outwardly visible, constitute the scope of natural history. However, when these parts are first presented to the eye in isolated form, we are concerned with the science of anatomy. The latter does not deal merely with outward form but with inner structure and thus quite properly makes use of the microscope.

When the organic body has been broken down in this way, so that its form is dissolved and its parts can be regarded as matter, the science of chemistry sooner or later steps in to give us new and wonderful information concerning the smallest ultimate parts and their combination.

When the destroyed creature is recreated from all these singly noted phenomena and is observed in its live and healthy state, we call the activity physiology.

Physiology is the mental operation performed in attempting to put together a whole from the animate and the inanimate, the known and the unknown, the complete and the incomplete, from perceptions and conclusions. Such a whole is simultaneously visible and invisible; its exterior must forever remain a whole to us, its interior forever a part; its actions and effects must remain eternally a mystery. Thus is it easy

---

[34] Used in the old sense of the term, as a synonym for preformation.

to see why physiology had to lag behind for so long and probably always will lag behind, namely, because man, though he feels his limitations, is seldom willing to acknowledge them.

Anatomy has been elevated to such a degree of exactness and precision that its well-defined facts in themselves comprise a kind of physiology.

Bodies are moved insofar as they have length, breadth, and weight, insofar as pressure and thrust act upon them, and insofar as they can be moved in some manner or other. For that reason men familiar with the laws of Nature have applied them also to organic bodies and their movements, not without advantage.

Chemists, too, have observed in detail the changes as well as the composition of the smallest ultimate parts, and their diligence of late as well as the accuracy of their methods justify their claims of having revealed the nature of the organic complex.

On the basis of all this, even aside from considerations not taken into account here, one can easily see how necessary it is to summon all our mental powers in a general striving for insight into all these mysteries, how necessary it is to use all mental and physical aids and to press every advantage in venturing to approach this never-ending work. Even a certain one-sidedness is not disadvantageous to the whole. Indeed, let each individual consider his own way best, if only he will smooth and clear it properly, so that those who follow him on the road may traverse it more easily and rapidly.

Recapitulation of the various sciences:

a. Knowledge of organic bodies, according to habitat and differences in form relationships.

Natural history.

b. Knowledge of material natures in general, as forces and in their place relationships.

Natural philosophy.

c. Knowledge of organic bodies, with reference to their inner and outer parts, without consideration of the living whole.

Anatomy.

d. Knowledge of the parts of an organic body insofar as it ceases to be organic, or insofar as its organization is regarded merely as substance-producing and substance-composed.

Chemistry.

e. Observation of the whole insofar as it is living and its life has a special physical power.

Zoonomy.

f. Consideration of the whole insofar as it lives and acts, and insofar as an immaterial power is at the basis of this life.

Physiology.

g. Consideration of form both in its parts and as a whole, the conformities and deviations, apart from all other consideration.

Morphology.

h. Consideration of the organic whole by visualizing and linking all these considerations through mental processes.

Consideration of Morphology in General

Morphology may be regarded both as an independent science and as an auxiliary physiological science. As a whole, it is based upon natural history, from which it extracts phenomena for its own purposes; it likewise rests on the anatomy of all bodies, and especially on zootomy.

Since its goal is merely to present and not to explain, it utilizes as little as possible of the other auxiliary sciences of physiology, although it does not disregard the energy and place relationships of physics, nor the material and compositional relationships of the chemist. A separate science only through definition, it is everywhere considered the handmaiden of physiology, co-ordinate with other auxiliary sciences of the latter.

In planning to establish morphology as a new science—to be sure, not with respect to the familiar phenomenon itself, but to the theory and method which gives the science its characteristic form, and indicates its place among the other sciences—we shall first show its connection with other related sciences and after that present the subject matter and its manner of exposition.

Morphology should include the theory of form, formation, and transformation of organic natures. It therefore belongs among various natural sciences, the purposes of which we shall now enumerate.

Natural history accepts the varied forms of organic life as known phenomena, yet scholars in the field must inevitably note that this diversity nevertheless exhibits a certain consonance, in part general and in part specific. These experts not only bring forward the bodies known to natural history but also arrange them—now in groups, now in series—

according to the characteristic qualities sought for and recognized, thereby making it possible to survey the enormous mass. Their work is twofold: in part, to continually hunt down new facts; in part, to arrange the facts more and more in accordance with Nature and the characteristic qualities mentioned, and to ban all arbitrariness insofar as possible.

Thus, natural history is restricted to the outward appearance of forms, regarding them as a whole; anatomy, however, aims at knowledge of their inner structure, at analysis of the human body as its most worthy subject—a subject which requires much auxiliary help and which is impossible without precise insight into its organization. Much has been done in the anatomy of other organized creatures, but the data are so scattered, usually so incomplete, and occasionally so falsely observed that the great mass is useless to the scientist.

Both related and unrelated sciences have been employed to enlarge and continue, to summarize and apply the data furnished by natural history and anatomy. Likewise, special points of view have been selected, always with the aim of meeting the need of general physiological perspective. In the process excellent spade work has been done for physiologists of the future, even though the methods used until now have been all too humanly one-sided.

From the physicist in the strict sense of the term, the science of organic life has been able to take only the general relationships of forces, their position and disposition in the given geographical location. The application of mechanical principles to organic creatures has only made us more aware of their perfection; one might almost say that the more perfect living creatures become, the less can mechanical principles be applied to them.

To the chemist, who annuls form and structure and is interested only in the properties of substances and their compositional relationships, morphology owes a great debt, a debt that will grow deeper as new discoveries make possible analyses and compounds of utmost delicacy, and therewith a closer approach to the infinitely delicate functioning of living organic bodies. We have already created anatomical physiology through exact observation of structure; and in time we can thus expect a physiochemical physiology. It would be desirable that the two sciences continue to go forward as if each were intent upon completing the whole task by itself.

However, since both sciences are analytic in nature—chemistry, in fact, resting solely upon analyses—it is natural that these two methods of describing and explaining organic bodies are not satisfactory to all scholars alike, many of whom have the tendency to proceed from a whole, developing the parts from it and immediately tracing them back again. For such a procedure the nature of organic bodies provides the finest opportunity, as the most perfected of such bodies make their appearance as a unit separate from all other entities. We ourselves are conscious of a state of complete physical well-being by sensing not parts of our bodies, but the body as a whole. All existence is possible only because bodies are organized, and are capable of being organized and preserved in their activity, solely through the state we call life. Thus, in an attempt to study the laws whereby life is given to organic nature, nothing could be more natural than to establish a zoonomy. Quite justifiably, a force was ascribed to this life for purposes of discourse; and this force could be, indeed had to be, assumed, because life as a whole expresses itself as a force that is not contained within any one part.

We cannot regard an organic nature as a unit for long, we cannot imagine ourselves as a unit for long, without being obliged to assume a double point of view, considering ourselves as an entity sometimes perceivable by the senses, and at other times recognizable only with the inner sense or noticed only by an effect.

Zoonomy thus embraces two parts not easily separated, namely, the material and the immaterial. Although the two cannot actually be separated, the worker in this field can nevertheless proceed from one or the other aspect and thus give ascendency to one or the other.

It is not merely that each of the sciences enumerated will require the services of an expert; each of their branches will demand the entire lifetime of an individual. An even greater difficulty arises from the fact that these sciences are carried on almost solely by physicians whose medical practice—much as it may contribute on the one hand to practical experience—deters them from broader study.

One can see how much preliminary work still remains to be done for the benefit of the individual who is subsequently to summarize all these observations from the standpoint of physiology, if he is to synthesize them and gain the insight befitting the grand subject insofar as it is possible for the human intellect. To gain our objective, we need purposeful activity—and of this there has been and continues to be no lack. In such activity, we would all proceed faster if we worked as specialists, not however one-sidedly setting our own opinions first, but joyfully acknowledging the merits of our co-workers.

After listing and pointing out the relationships of the various auxiliary sciences of physiology, it is time to legitimize morphology as a separate science.

In fact, that is the way it is regarded. It justifies itself as a separate science by taking as its chief subject one that is treated only occasionally and incidentally by other sciences, by collecting data scattered in others, and by selecting a vantage point from which natural phenomena can readily and easily be observed. Morphology has a great advantage in concerning itself with generally recognized elements, in not being controversial and therefore not being forced to make a place for itself at the expense of another science, in occupying itself with phenomena of the utmost significance, in employing in its summaries intellectual operations so adapted and pleasing to human nature that even unsuccessful investigations may still give both benefit and pleasure.

1. *Methods of Approach*

To organize my undertaking is an extensive and difficult task.

To achieve good organization requires exact knowledge of the subjects under consideration.

Attention to their characters, in other words, to deviations and conformances.

Such attention requires far more than sense perception and memory.

Insight into what is significant, and evaluation of it.

Effort of the human intellect to round out and make a whole of the subject treated.

Impatience of scholars, preventing them from being sufficiently prepared.

Hastiness in drawing conclusions.

Such hastiness not always to be condemned.

Experiences of various historical periods.

The earlier ones less complete.

No one intending to acquire scientific knowledge can know at the very outset that again and again he will have to set his sights higher with respect to viewpoint and way of thinking.

Those who were occupied with the sciences felt this need only gradually.

In this day and age, when so many generalities are discussed, the botanical gardener, who is little more than a day-laborer, encounters the most difficult questions, and since he is not acquainted with the theoretical knowledge that might provide the answers, he either has to get them from someone else or remain in a state of astonished confusion.

One, therefore, does well to be prepared at the very start for serious questions and sober answers.

Desiring to set one's mind somewhat at rest on this score and to procure a higher vantage ground, one should tell one's self that no one asks a question of Nature that he cannot himself answer, for the answer is inherent in the question, in the feeling that the point can be discussed and pondered.

To be sure, the questions vary according to the different types of humans.

To orient ourselves somewhat among these various types, let us divide them thus into four spheres: utilizers, fact-finders, contemplators, and comprehenders.

1. The utilizers, advocates and seekers of things practical, are the first to plow the field of science, metaphorically speaking, and they aim at practical results. Self-confidence derived from experience gives them assurance; necessity gives them a certain breadth.

2. Fact-finders, those who crave knowledge for its own sake, require a calm, disinterested gaze, an inquisitive unrest, a clear mind. They are in constant contact with the first group, but work out the results from the scientific point of view exclusively.

3. The contemplators are somewhat more original, for the mere increase of knowledge unwittingly fosters interpretation and crosses over into it. Even the fact-finders, however much they may make the sign of the crucifix at the very thought of imagination, before they realize it are compelled to call upon this selfsame power for assistance.

4. The comprehenders—in a deeper sense they might be called creators—are original in the highest sense of the term. By proceeding from ideas, they simultaneously express the unity of the whole, and it is almost the obligation of Nature to conform to the ideas.

Simile of roads.

Illustration of the aquaduct, to distinguish between the fantastic and the ideal.

Illustration of the dramatic poet.

Creative imagination with all possible realism.

In all scientific effort one must make clear to himself that he will move in these four spheres.

One must ever be conscious of the sphere he is working in at the moment.

And one must have the inclination to move in one as freely and easily as in another.

The objective and subjective in one's exposition is thus recognized and separated in advance, and in this way one can at least hope to inspire a degree of confidence.

II. *Genetic Treatment*

In our exposition, obviously, we shall tarry for the most part within the limits of the second and third spheres, consciously moving from one into another.

By instinct, the fact-finders as a rule take refuge with the interpreters, although through false teleological reasoning they often return in theoretical situations to the level of the utilizers—and among these we must reckon all natural scientists to the glory of God.

Genetic treatment provides a point where the meeting place of the two regions can be made clear and be utilized.

When I see before me something which has already taken shape, inquire about its origin and trace back the process as far as I can follow, I become aware of a series of stages. Naturally, these cannot be observed side by side with the physical eye but must be pictured mentally as a certain ideal whole.

Inclined at first to postulate certain stages, I am finally compelled, since Nature never proceeds by skips and jumps, to regard the sequence of uninterrupted activity as a whole, annulling individual details so as not to destroy the total impression.

Rough plotting of factors.

Attempt at a more precise plotting.

Attempt at finding several intermediate points.

Reflecting upon the results of these attempts, one sees that eventually fact-finding must be terminated, contemplation of a growing thing must begin, and ultimately an idea must be expressed.

Example of a city as a work of man.

Example from the metamorphosis of insects as a work of Nature.

Theory of the metamorphosis of plants in its total significance.

III. *Organic Unity*

Identity of the parts in their most diverse forms.

Important questions entering into the discussion:

Whether what is present has developed from the seed?

Whether definite beginnings continue to be formed and transformed according to law?

The atomistic point of view has a certain similarity to the general point of view.

To a certain mentality.

It cannot be dispensed with entirely in natural science.

But it becomes an obstacle if followed slavishly throughout.

Certain minds cannot rid themselves of the atomistic point of view.

Dynamic points of view.

Their initial difficulties.

Their ultimate advantages. Several contrasting examples of both points of view.

The dynamic point of view is to be used provisionally in our discussion.

It must justify itself in use.

In plant study we assume a beginning point of life that eternally reproduces itself.

And in very few cases is the reproduction achieved by mere repetition.

Furthermore, in the more perfect organisms it is done through progressive development and transformation of the basic organ into more and more elaborate and effective organs, to achieve ultimately the highest point of organic activity: individuation and release of individuals from the organic whole through generation and birth.

Supreme view of organic unity.

IV. *Organic Duality*

Up to this point we have regarded the plant as a unit.

We can recognize its empirical unity with our eyes.

Originating from the combination of many different parts of greatest diversity, it forms an apparent individuum.

A semi-annual completed plant plucked apart.

Ideal unity:

If these different parts are visualized as having originated from an ideal archetype, gradually developing by various steps.

These ideal archetypal bodies, even though we may picture them in our minds as simply as possible, we must nevertheless imagine as disunited in their interiors, for no third developing body can be imagined without previous division of the original body.

These ideal archetypal bodies, which already have an inherent tendency toward duality, we shall allow to rest for the present in the womb of Nature.

We shall only remark here that the atomistic and dynamic concepts are opposed at the very outset to developmental and formational procedures.

Brief general discussion of dualism in Nature.

Transition to the plant.

Although in an organic body, this transition almost partakes of the character of physics.

Origin of the root and of the leaf.

They are united by origin; indeed, the one cannot be imagined without the other.

They are also by origin opposed to each other.

We answer the question, why the root embryo develops downward and the leaf embryo upward, by saying that they are opposed, in keeping with the general dualism of Nature, which here becomes specific.

However, something more detailed remains to be said about environment.

A plant, like every other natural entity, cannot be imagined without an environment.

It needs a base of existence, as a fastening and for the chief nourishment of the mass.

It requires air and light for varied development, more refined nourishment for elaboration.

We find that the roots require moisture and darkness to develop; the leaf requires light and aridity.

Thus, from beginning to end these needs are opposed.

With conditions of moisture and darkness, to a certain extent even the latter alone, the root can develop at any node and even at many other points of the plant body.

The leaf embryo can develop at any point on the plant as soon as light and dryness act upon it.

Examples.

Chief differences between the root and leaf embryo.

The former always remains simple.

It is merely a combination without diversity.

The leaf embryo, on the other hand, develops most diversely, and step by step approaches perfection.

Light and dryness foster elaboration.

Moisture and darkness retard it.

Certain plants, vines in particular, which in spite of light and dryness develop a quasi-root, have much that is watery in their constitution, together with a certain tenacity and irritability.

If we can imagine such a plant with a general schism at its beginning, we shall be able to find such division again in its parts.

We shall find the division again in the upper and lower surface of the leaf.

In the splint, forming wood on the inside and rind on the outside, until at length we reach the climax of organic duality in the division into two sexes.

Later Studies and Collections[35]

My theory of metamorphosis could not possibly be treated adequately in a single, independent work; it could only be advanced as an ideal, as a gauge against which organic structures might be held and measured. In attempting to penetrate deeper into the realm of plants, then, the first and most natural thing for me was to try to evolve a detailed conception of the various phenomena and of their origin. Since it was my intention also to continue the written presentation of the study I had begun, discussing in detail what I had merely touched on in the original essay, I collected examples of formation, transformation, and malformation, examples which Nature provides so liberally. I had drawings, paintings, and engravings made of those that seemed particularly illuminating; and by appending these illustrations of striking phenomena to the various sections as I wrote them, I prepared the sequel to my first work.

Through stimulating association with Batsch,[36] the importance of relationships between plant families had gradually become clear to me, and now Usteri's edition of Jussieu's[37] work proved extremely useful. The acotyledons I dropped, studying them only when they approached definitive form. However, it could scarcely escape me that the most rapid means of making a survey was afforded by a study of the monocotyledons. Indeed, by the simplicity of their organs, these plants openly display the secrets of Nature, pointing forward to the more developed phanerogams and backward to the mysterious cryptogams as well.

Active as I was, driven hither and yon by heterogeneous pursuits, diversions, and passions, I was content to rearrange in private the knowledge I had acquired, and to use it for my own purposes. Contentedly, without publishing my ideas, I pursued the whimsicalities of Nature. The great labors of Humboldt, detailed works of scientists of all nations, afforded material enough for quiet study. Eventually, the desire reawoke

---

[35] Begun in September, 1817, these notes were first published in *Natural Science in General; Morphology in Particular*, Vol. I, No. 2 (1820).

[36] August Karl Batsch, 1761–1802, professor, Jena. See also pp. 76, 155, 183, 189, 196, 217.

[37] For references to Usteri and Jussieu, see p. 182.

to carry on these studies more actively, but just when my dreams were approaching reality, the engravings were lost and I had neither inclination nor courage to do them over again. Meanwhile, my method of interpretation had captured young minds, developing more vigorously and consistently than I had deemed possible; and from then on I sanctioned all excuses that fostered my indolence.

After these many years, when I contemplate what is left of my efforts and re-examine the materials that remain—plants dried or otherwise preserved, drawings, engravings, marginal notes to my first essay, collectanea, excerpts from books and reviews, numerous publications—I can clearly see that the goal I had in mind must inevitably remain unattainable for a man in my position, for a man of my temperament and methods of study. What I had undertaken to do was nothing less than to present to the physical eye, step by step, a detailed, graphic, orderly version of what I had previously presented to the inner eye conceptually and in words alone, and to demonstrate to the exterior senses that the seed of this concept might easily and happily develop into a botanical tree of knowledge whose branches might shade the entire world.

Today I am in no wise saddened by the thought that this project kept eluding me, for since then science has taken a step forward, and capable men now find abundant means at hand for advancing it still further. Draughtsmen, artists, engravers—so well instructed and informed are they that they might be regarded as botanists themselves! Indeed, anyone who desires to copy, to re-create, must necessarily understand the subject and must look deeply into it; otherwise he will get into his picture only an outward resemblance and not Nature herself. Such men are needed if paintbrush, etching needle, and graving tool are to give true account of delicate nuances as form merges into form. They, especially, must perceive, first with the intellectual eye, in an evolving organ the form that can be expected and must eventually follow. They must see the norm at work in the abnormal; they must discover the rule operating in the exception.

Thus, if a discerning, energetic, and enterprising man were to plunge in and definitively arrange, classify, and organize all pertinent material, we should have good hope of satisfactorily accomplishing a task that formerly would have seemed impossible.

To be sure, lest once again we spoil a good thing, we ought to proceed from the actual, sound, physiologically pure metamorphosis and only then explain the pathological, the uncertain progressions and regressions of Nature—true malformations of plants—to put an end in this way to the obstructive mode of procedures resulting from the use of the term metamorphosis solely in reference to irregular formations and to malformations. In the latter case, however, the work of our worthy Jäger[38] will prove valuable and stimulating for purposes of orientation and reference. In fact, this faithful, industrious observer might have anticipated all my own desires and might have been able to execute the work which I have just outlined, had it been his intention to study healthy as well as diseased plants.

Here may be the proper place for a few remarks I wrote down on first becoming acquainted with this highly stimulating work of Jäger.

In the plant world, what is completely normal is correctly called healthy and physiologically pure. But on the other hand, the abnormal should not necessarily be regarded as diseased or pathological. At the most, we might list the monstrous in that column. In many cases, therefore, we do ill to speak of "failures" and "deficiencies," which would indicate that something is missing, for it might just as well be a case of superfluity, or of development occurring without balance or contrary to it. Also the words misdevelopment, malformation, crippling, and stunting should be used with care, for even though Nature operates with greatest freedom in this realm, she nevertheless may not depart from the fundamental laws of her being.

Nature fashions normally when she subjects innumerable particulars to a rule, defines and delimits them. Conversely, the phenomena are abnormal when the particulars carry the day, emerging in an arbitrary, indeed apparently accidental way. However, since both are closely related, and since the same spirit animates the regular and the irregular as well, an oscillation between the normal and the abnormal occurs, formation and transformation forever alternating, so that the abnormal seems to become normal, and the normal abnormal.

The shape of a plant can be changed or lost, and yet we should not care to allude to this as malformation. The centifolia[39] is not malformed, although it is indeed abnormal. However, the term malformed is rightfully applied to the perfoliate rose,[40] for here the lovely rose form is dissolved, and the lawful limits are exceeded.

---

[38] George Friedrich Jäger, 1785–1866, professor in the classical high school of Stuttgart. The work referred to is *Observations on Malformations in Plants,* a contribution to the history and theory of misdevelopments of organic bodies. Goethe valued this book highly and made frequent use of it. It had appeared in 1814, but he did not occupy himself with it until 1816.
[39] The double rose.
[40] See par. 103 and par. 104, p. 73.

All double flowers we classify as abnormal, yet it is perhaps deserving of mention that such flowers possess heightened beauty and fragrance. Nature oversteps the boundary she has set, but attains thereby a different kind of perfection; thus we would do well to defer the use of negative terminology as long as possible. The ancients said *teras, prodigium, monstrum,* a miracle sign, fraught with meaning, worthy of all attention; and, with this in mind, Linné very felicitously named his "peloria."

I wish that men's minds might be permeated with the truth that a complete view can never be attained unless the normal and the abnormal are regarded as constantly fluctuating and operating in a direction toward one another. Here we shall insert a few pertinent details on the subject.

When Jäger* speaks of malformation of the root, we also call to mind its healthy metamorphosis. Above all, we are reminded of the root's identity with trunk and branch. We once saw a highway being built over an old beech-covered hillside which had to be cut off sharply to level off the road. Hardly had the age-old roots glimpsed the light of day, hardly had they had a taste of the quickening heavenly element, when instantaneously they took on the appearance, all green, of young bushes. It was striking to see, and yet similar things maybe observed daily; indeed, every gardener finds his attention drawn to the important work of propagation, compelled as he is to constantly clear the ground of runaway roots which are forcing up twig upon twig in the manner of branches.

On examining this change in the shape of the root, we find that the usual threadlike formation can change in diverse ways, especially by protuberation. The napiform shapes are familiar to everyone, likewise the bulbiferous. The latter are protuberant roots, complete in themselves, with shoot upon shoot distributed over the surface. An example of this is our edible potato, the diverse propagation methods of which rest in the identity of all parts. Stem and branch strike root as soon as they are placed in the earth, and so on, ad infinitum. We were once witness to a charming instance of this. In a vegetable garden, a single potato vine was growing among the cabbages. Remaining unnoticed, its branches grew downward and kept moist in the shade of leafy plants. In the fall the stems emerged swollen into tiny elongated potatoes, at the top of which a tiny crown of leaves peeped out.

We are also familiar with protuberant stems above the ground, in kohlrabi as a preparatory organ from which the blossom grows directly, in pineapple as a completely developed and fructified organ.

---

* Jäger *(op. cit.)*, p. 7.

With improved nourishment, a stemless plant develops a considerable stalk. Amid dry rocks, on barren sunny limestone, *Carlina*[41] emerges completely acaulescent, but if it finds its way into earth that is even the least bit loose, it lifts itself immediately; and in good garden soil it is no longer to be recognized, for it has acquired a tall stem and then takes the name *Carlina acaulis, caulescens*. Thus Nature forces us to alter our classifications and to defer to her independent working and reworking of things. In the same way, we must give the science of botany credit for constantly altering its terminology in the direction of precision and flexibility. Of this, incidentally, we have striking examples in the last issues of Curtis' *Botanical Magazine*.[42]

If the stem is forked, if there is a change in the number of stem angles, if a broadening takes place*, these three phenomena provide evidence once again that in organic formations, several identical forms can and must develop, in, with, beside, and after one another. They indicate multiplicity in unity.

Every leaf, every bud, is in itself entitled to become a tree. It is only the superior health of the stem that prevents the leaves and buds from attaining that state. We cannot repeat often enough that each organization unites various active parts. In the present instance, we observe that the stem is usually round or at least may be regarded as round within. And it is precisely the roundness, in its singleness, that holds asunder the individual parts such as leaves and buds, and allows them to ascend in ordered sequence to regular development until they blossom and bear fruit. But if such a plant entelechy is checked, if not indeed done away with completely, the middle loses its governing power, the periphery contracts, and each individual, ambitious part now exercises its own special right.

Such is frequently the case in the crown imperial;[43] a flattened and greatly broadened stem appears to consist of thin reed bars crowded in groovelike formation; and the same flattening occurs also in trees. The ash, particularly, is subject to this deviation; however, in its case the periphery is not immediately pressed flat. The branch has a wedge-shaped appearance and loses its regular growth first at the sharp end, while the wood formation still continues at the broader top end. The

---

[41] A type of thistle.
[42] William Curtis, 1746–1799, English botanist, founded his *Botanical Magazine* in 1774.
* Jäger {*op. cit.*}, pp. 9–20.
[43] *Fritillaria imperialis*.

lower, narrower part is therefore first to thin out, contract, and lag behind, while the upper part continues to grow vigorously and still produces complete branches. In spite of all this, welded as it is to the arrested plant part, it bows down. However, to compensate for the humiliation, it acquires the wondrously regular form of a bishop's crooked staff, thus becoming a creative symbol for the artists.

This broadening out is also remarkable in that we can quite properly call it a prolepsis,[44] for we find here a precipitate eagerness to force out and to mold the bud, blossom, and fruit. And far more perfect flowers appear on the flattened stem of the crown imperial, as well as of the monkshood, than would ever have been produced on the healthy stem. The bishop's crosier, which is produced by the flattened ash branch, ends in innumerable gemmae, which do not, however, develop further but remain in their dessicated state, a dead conclusion to their stunted growth.

Such a flattening out is natural to *Celosia cristata*. Its cockscomb develops innumerable sterile blossoms, but a few of these, close to the stem, produce seeds which possess to a certain degree the quality of the mother plant. In general, we find again and again that deformations return to the pattern, that Nature has no rule to which she might not make an exception, and makes no exception which she might not again turn back into a rule.

If the division of the leaves* were to be regarded as a misdevelopment in each instance, the true value of the observation would be lessened. When leaves divide, or rather when they advance from their original state to diversity, they are striving toward greater perfection, in the sense that each leaf has the intention of becoming a branch, and each branch a tree. All classes, families, and orders have the right to struggle toward such a goal.

Among the ferns we find splendidly plumed leaves. How powerfully the palm manages to wrest itself from the usual monophyllous state of the monocotyledons! What plant lover is unfamiliar with the development of the date palm, which may appropriately be cited here as an example of such striving toward diversity from its first expansion onward. Its first leaf is as simple as that of Indian corn; then it is divided into two, and the fact that it is not mere laceration taking place here is shown by the existence of a small vegetable seam at the base of the incision,

---

[44] See par. 109, pp. 74–75.
* Jäger {*op. cit.*}, p. 30.

welding duality into unity. Further division then takes place and at the same time the rib pushes itself ahead, forming in the process a much incised branch.

I might cite as an example the fan palm in the Padua botanical garden,[45] throughout its development up to the blossoming stage. For it is a healthy, organic, necessary, and prearranged metamorphosis, without interruption, disturbance, or false steps, which is in operation here. Especially remarkable are the seams which link the radially-divergent, lance-shaped leaves, and thus give rise to a perfect fan shape. Further examples of that kind are to be urgently recommended for future illustrated essays. Most remarkable of all are the branchlike leaves of leguminous plants; their wondrous, manifold development and their sensitiveness point toward higher attributes, vigorously and wholesomely manifested in the root, cortex, stem, blossoms, husks, and fruit.

This division of the leaves is subject to a law which is easy to demonstrate visually but difficult to express in words. The simple leaf divides at both sides of the leaf stalk's base, becoming threefold, and the uppermost of these leaves divides in its turn at the base, with the result that another threefold leaf originates. Thus the leaf as a whole must be regarded as fivefold. At the same time it is already evident that the two lower leaves have the tendency to divide on one side, on the margin pointing downward; and this does indeed occur, giving rise to a sevenfold leaf. This division continues; the upward-directed margin becomes incised and divided, giving rise to a ninefold divided leaf; and so on.

This phenomenon is striking in *Aegopodium podagraria,* of which the amateur botanist can easily obtain the complete collection. In so doing, he will notice that the manifold division is much more frequent in shady and moist places than in dry and sunny areas.

The regression of such division may also occur. Probably the most striking example is an acacia from New Holland, which emerges from the seed with pinnate leaves and gradually transforms them into onefold lance-shaped leaves. The process is effected by the flattening of the lower end of the leaf stalk and its gradual engulfment of the pinnate portions originally present at the top. Here we perceive that Nature has the choice of going backward or forward in the selfsame way.

In *Bryophyllum calycinum,* which as a whole is extremely remarkable, we have observed that the approximately half-year-old plant, after having diversified its leaves into three parts, again produced simple leaves in

---

[45] See p. 161.

winter and that this singleness continued up to ten leaf-pairs, at which point, in midsummer when the plant was a year old, the threefold division again occurred. We are now waiting to see how this plant, which is now pushing out its leaves in a fivefold division, will behave henceforward.

In the category of abnormal growth, we must also include the plants that have been etiolated by design or accident. When, inconsistently with their nature, they are deprived of light, they behave in part like roots running below the earth and in part like stolons creeping along above the earth. In the first case they remain white and continue toward their goal; in the latter case they produce buds to be sure, but these buds do not advance toward perfection—no metamorphosis takes place. Larger plants stop growing. A number of individual details still remain for future treatment.

Blanching is usually an intentional etiolation of the leaves, effected by tying certain plants together, thereby causing the inner ones, which are deprived of light and air, to assume unnatural attributes.

In form, there is distention of the midrib as well as proportionate distention of its ramification; the leaf remains smaller, because the intermediate spaces of the ramification are not filled out.

In color the leaf remains white, deprived as it is of the operative effect of light. In flavor it remains sweet, for it is precisely the operation which enlarges the plant and colors it green that also appears to foster bitterness. Likewise, the fibre remains delicate and everything serves to make it palatable.

It frequently occurs that plants will sprout in the cellar. When this happens with kohlrabi, the offshoots are tender white stems accompanied by a few leaf tips, as delicious as asparagus.

In southern Spain palm tops are blanched by tying them together; the branches keep on growing because the germinating power cannot be stopped, but they remain white. They are carried by the highest clergy on Palm Sunday.[46] In the Sistine Chapel, one can see the Pope and the cardinals adorned with them.

Fruit within Fruit.* In the autumn of 1817 after the double poppies had stopped blooming, little poppy heads were found containing complete, still tinier poppy heads. The stigma of the inner one at times almost touched that of the outer one, and at other times it remained apart, closer

---

[46] As symbols of divine purity.
* See Jäger {*op. cit.*}, pp. 218 ff., esp. p. 220.

to the base. The seeds of several were preserved, but it was impossible to observe whether this character was transmitted.

In 1817 a remarkable ear of corn from both sides of which ten smaller ears were growing, was found in a field belonging to Adam Lorenz, a farmer from Niederhausen an der Nahe, in the vicinity of Kreuznach. A drawing of it was sent to me.

Here I might cite other particulars from my notes to Jäger's book, but instead of working further on these subjects, fragmentarily, inadequately, and without illustrations, I shall merely refer my readers to a man whose achievements mark him as one who may eventually solve these riddles, a man who might gently prod us and properly direct us to the right road toward our goal, a road upon which faithful and brilliant observers are at present wavering back and forth, half astray. That this man is our esteemed friend President von Esenbeck,[47] will be cheerfully acknowledged by every German naturalist the moment his name is mentioned. He first proved his skill in experiments with almost invisible forms, perceptible only to the sharpest eye. Then, working with double forms of life the two parts of which were developing divergently, and later with completely determined species, he demonstrated a method which might be used in the classification of species, to show how one develops from the other in turn. Intellect, knowledge, talent, and position—all these demand and justify that he be acknowledged here as a collaborator.

Let him share the triumph of physiological metamorphosis with me. Let him demonstrate the process at the point where the whole is divided, separated, and transformed into families, the families into genera, and the genera into species. Nature goes on and on with this activity; she cannot stop or rest. But neither can she preserve and maintain all that she produces. For do we not have the most conclusive remains of creatures that were unable to perpetuate themselves by energetic propagation! On the other hand, there are those plants which, as they develop from the seed, constantly deviate from the standard and constantly alter the relation of the parts to one another. Our faithful, painstaking observer has already provided us with information on some of these, and without doubt he will report on more of them as time goes on.

We are again convinced of the importance of all this observation when we give a final glance at the place where families become distinct

---

[47] Christian Gottfried Nees von Esenbeck, 1776–1858; president of the Imperial Leopold-Caroline German Academy of Natural Scientists.

from families: here formation and malformation already brush against each other. Who could take it amiss if we were to classify orchids as monstrous *Liliaceae?*

Pollination, Volatilization, and Exudation[48]

WE WOULD NOT be shooting wide of the mark, were we to regard these three closely related phenomena, which often make their appearance simultaneously and linked together, as symptoms of a ceaselessly advancing organization, hurrying from life to life, indeed through annihilation to life. Let me briefly summarize here what I have observed in this regard, and the conclusions I have reached.

It is approximately sixteen years ago that Professor Schelver,[49] curator of the Grand-ducal Immediate Botanical Garden under my direction, disclosed to me in strictest confidence, in that very garden on the same paths where I still take my walks, that he had long had his doubts about the theory ascribing two sexes to plants as well as to animals and that he was now fully convinced of its untenability.

In my nature studies I had religiously accepted the dogma of sexuality in plants and was, therefore, taken aback now to hear a concept directly opposed to my own. Yet I could not consider the new theory wholly heretical, for from the account given by this ingenious man I could draw the conclusion that the pollination theory was a natural consequence of the theory of metamorphosis which seemed so significant to me.

Now the doubts that had been raised from time to time about the sexual system came to my mind, and the ideas that I myself had had on the subject came to life again. The new viewpoint supported certain views of Nature that now seemed clearer and more significant to me; and accustomed as I had always been to preserve complete flexibility in my application of metamorphosis, I likewise found this viewpoint not uncomfortable, although at the same time I could not immediately relinquish the other.

No one familiar with the situation of our botanical science at that time will blame me for imploring Schelver not to let his thoughts get abroad. It was to be foreseen that his theory would get a most unfriendly reception and that the theory of metamorphosis, which as it was had

---

[48] Written during the summer of 1820, published in *Natural Science in General; Morphology in Particular,* Vol. I, No. 3 (1820).

[49] Friedrich Joseph Schelver, 1778–1832, for a time university professor in Jena. For further references to Schelver, see pp. 119, 189, 199, 241.

found no acceptance, would be banished from the boundaries of science for a long time to come. Our own academic standing also made such secrecy advisable, and I am grateful to him down to the present moment for linking his conviction to mine, for not speaking a word about it for as long as he remained with us.

In the course of time, however, the sciences, like everything else, had undergone a good deal of change. One new theory appeared after another, and even bolder ideas had been expressed when Schelver finally stepped forward with his daring new idea.[50] Thus it was only to be expected that his theory would have to remain an open secret to the world for somewhat longer. Opponents appeared and he was thrust back protesting from the threshold of the temple of science. And he fared no better when he could no longer refrain from self-defense.

He and his crotchet were brushed aside and silenced. However, it is a peculiar characteristic of the times that a seed, once sown, will immediately strike root somewhere. Receptivity is great, the true and the false sprout and grow up together in vigorous confusion.

Now his brilliant theory takes on substance through Henschel's[51] momentous study; it is earnestly demanding its place in science, although one cannot yet foretell how that place will be found. However, interest in the theory is already astir. Reviewers, instead of preaching and scolding against it as before, confess that they have been converted, and now we can only wait to see how things will develop further.

Just as there are extremists everywhere at present, among liberals and monarchists as well, so Schelver was an extremist in the theory of metamorphosis, breaking down the last barrier that kept it prisoner within the limits formerly prescribed for it.

His thesis and his defense of it at least can never be excluded from the history of botany; as brilliant exposition, or even as mere hypothesis, it merits our attention and interest.

In general, scientists should be trained to follow the reasoning processes of others. For me, as a dramatist, this was not a difficult task; for a dogmatist, it is admittedly a hard assignment.

Schelver proceeds literally from the concept of healthy and regulated metamorphosis, which holds that plant life, rooted in the earth, strug-

---

[50] Schelver's book, *Critique of the Theory of Sexuality in Plants*, appeared in 1812.

[51] August Wilhelm Eduard Theodor Henschel, 1790–1856, physician in Breslau, who published a work, *On the Sexuality of Plants*, in 1820.

gling toward light and air, is forever raising itself by its own bootstraps and developing step by step, scattering about even the last seed by its own force and power. The sexual system, on the other hand, requires for this final act an external agent, which is conceived of as the counterpart of the flower itself, exerting influence and being influenced—with, beside, or even apart from the flower.

Schelver pursues the tranquil course of metamorphosis, which, in advancing, is refined to such a degree that it gradually leaves behind all that is material, insignificant, and base, permitting what is higher, incorporeal, and better to emerge in greater freedom. Why, then, should not this latter type of pollination also be a liberation from burdensome matter, allowing the inherent abundance in the heart of the plant, through the energy of primary force, to proceed toward endless propagation.

Let us take as an example the sago palm, felled for the powder which is produced throughout its trunk as the tree approaches the flowering period, and which is kneaded into a nourishing food preparation. As soon as the flowering stops, the powder likewise disappears.

We know that the blossoming barberry bush diffuses a strange odor that may cause wheat fields in the vicinity to become unproductive. An unusual quality may be hidden within this plant, as we may indeed infer from the sensitiveness of its anthers. It diffuses its pollen insufficiently during the flowering period; thus, we later find bits of pollen emerging from leaves which may even develop in the manner of calyx and corolla to form the most magnificent cryptogam. This usually occurs with the leaves of branches from the previous year, which were capable of producing flowers and fruit. Fresh leaves and shoots of the current year are seldom productive in this manner.

Late in the season the under sides of centifolia leaves are covered by a dust that can be removed only with difficulty. On the other hand, the upper sides are streaked with pale places, from which it may clearly be seen that the undersides are tabetic. Now if it were found that the same phenomenon does not occur with simple roses, which fully carry out the act of pollination, then it might be considered quite natural for the centifolia, whose pollination organs are more or less abolished and transformed into petals.

Grain rust furnishes an example of delayed pollination terminating in nothingness. Through what irregularity of growth does a plant sink into a condition where, instead of joyfully and vigorously developing a numerous progeny, it tarries on a lower step and finally executes the pollination act perniciously?

Most striking is the case when Indian corn is attacked by this disease. The kernels swell up, making the ears large and misshapen: the enormous quantities of black powder which they contain are evidence of the great nutritive powers concentrated in healthy corn—nutritive powers which here disintegrate to nothingness.

We see that anther pollen, which cannot be denied a certain organization, might well be assigned to the realm of mushrooms and fungi. Abnormal pollination has already been accepted there. Now let us grant the same privileges to the regular type.

There is no doubt that all organic pollination proceeds according to a certain rule and order. An unopened champignon, placed with cut stem on white paper, will soon unfold and pollinate this clean surface, in a regular pattern, and to such an extent that the entire structure of the inner and lower folds is clearly delineated. From this it becomes evident that the pollination does not occur haphazardly; instead, each fold contributes its share in predetermined sequence.

Also among insects a similar and eventually destructive pollination takes place. In autumn we can observe that flies fasten themselves on windows, remain motionless for a time, and then gradually eject a white pollen. The chief source of this natural phenomenon seems to be situated at the point where the midriff is joined to the hind part. The pollination takes place gradually and continues for some time after the creature's death. We can estimate the force with which this matter is discharged, from the fact that it is forced out from the center a half inch on both sides, so that the limbus appearing on both sides of the creature amounts to more than one Rhenish inch.

Although this pollination is most common and most pronounced toward the side, I have nevertheless noticed that sometimes it is exuded from the front parts, so that the creature is almost, if not completely, surrounded by *an expanse of pollen.**

If we can accustom ourselves to various intellectual approaches, it will never lead us to the equivocal in our observations of Nature. We may interpret things as we will, they will always remain the same for us and for others who follow.

---

*Renewed attention to the pollination act of dead flies permits me to conjecture that actually it is the rear part chiefly that ejects the pollen, and this with increasing elasticity. The pollination begins about one day after death; the fly remains firmly fixed to the window pane, and the fine pollen leaves its traces for four or five days, at always greater distances until the limbus which is forming measures an inch in transverse section. The insect does not fall from the pane except through external force or contact.

ON MORPHOLOGY 109

For the instruction of young persons and ladies this new pollination theory will be extremely welcome and suitable. In the past the teacher of botany has been placed in a most embarrassing position, and when innocent young souls took textbook in hand to advance their studies in private, they were unable to conceal their outraged moral feelings. Eternal nuptials going on and on, with the monogamy basic to our morals, laws, and religion disintegrating into loose concupiscence—these must remain forever intolerable to the pure-minded!

The accusation is often made, and not entirely without justification, that philologians frequently spend more effort than is proper on indecorous and frivolous passages, in partial compensation for the disagreeable dryness of their work. Similarly, one might point an accusing finger at the naturalists when they take as ribald a delight in Mother Nature as in the goddess Baubo herself—just because they have discovered a few little weaknesses in the good mother. Indeed, we recall having seen arabesques in which the sexual relations within a flower calyx were represented, in the manner of the ancients, in an extremely graphic way.

Incidentally, the botanist held no great brief for the old system. He believed in it as he would in any other dogma; that is to say, he allowed it to stand without making detailed inquiries into its roots and origins. And since it was customary to juggle the nomenclature about, no changes were necessary in the new system. Anther and pistil kept their old names, except that a specific sexual function was now attributed to them.

Turning now to exudation, we find that it also occurs normally and abnormally. The nectaries—the name should be interpreted literally—and the drops they are releasing, appear to be highly significant organs, related to the pollination organs. Indeed, in certain cases they perform the same function, be that function what it may.

During the honeydew, which was this year unusually pronounced, an attentive nature lover observed the following:

In the last days of June[52] we had a honeydew of unusual intensity. Over a period of four weeks the weather had been cool—some days were actually cold—with intermittent showers, local for the most part. General rains were less frequent. Then came fair weather with very warm sunshine.

Soon the honeydew could be observed on various plants and trees. Although after a few days the sight had become familiar enough to me and others, nevertheless one occurrence did astonish me. Along the moat, under an avenue of ancient lindens about to burst into bloom, I became

---

[52] 1820, in Jena.

aware of moisture that had settled, as from a dense mist, over the blocks, consisting mainly of flinty and clay slate, with which the street had recently been paved. When I returned after an hour and saw that the spots had not disappeared in spite of the bright sunshine, I investigated some of the stones more closely and found that the wet spots were sticky. Moreover, there were blocks that were completely covered, and the ones of flinty slate, particularly, looked as if they had been enameled black. Now it struck me that they lay in areas whose peripheries reached as far as the trees extended their branches, showing clearly that the moisture must have emanated from them. Closer examination then revealed that the leaves were glistening too, and the source of the dripping was thus verified.

Upon visiting a garden I found a tree, a Reine Claude, on which the moisture was so heavy that almost every leaf tip supported a drop of honey-like consistency, too thick to fall. Yet there were isolated places where drops had fallen from the higher leaves to those below. Such drops were always bright yellow, whereas the ones which remained on the leaves were always smudged with gray-black.

In the meantime the plant lice had gathered by thousands on the under side—those on the upper surface were usually stuck fast—and there were quantities of the discarded empty shells. The insects may have metamorphosed here or have perished; but in either case we may certainly assume that the honeydew was not produced by them, for I came upon lindens with leaves which looked as if they had been lacquered, but upon which there were neither plant lice nor carcasses.

This moisture comes from the plant itself, for directly beside one such linden stood another, undoubtedly a later-blooming one, without any moisture whatever. Likewise the lindens already in bloom exhibited no honeydew, or only very little.

On the fifth of July, after several light rains of short duration, the bees which had been buzzing about those lindens not yet in bloom, had begun to congregate on the leaves and were sucking up the honeydew. Perhaps the rain had washed the unpalatable part away and the little creatures were finding the remainder suitable for their purposes. This conjecture deserves consideration on the score that the bees were not observed on all the lindens that were covered with honeydew.

It should be noted also that a white currant bush was covered with this substance, whereas the red one directly beside it was not.

After so many observations, some interpretations may now be ventured. In May the branches and leaves had developed to a considerable size; June was wet and cold. A disturbed growth must inevitably fol-

ON MORPHOLOGY 111

low, for although the juices circulating in the roots, trunk, and branches were absorbed as much as possible by the branches and leaves, nevertheless the transpiration of the leaves could not proceed in the proper manner because of the cold and damp outer air, this long-sustained condition causing general retardation. Suddenly there came dry warm days, with temperatures of 20 to 26 degrees.

Now the trees and plants, containing abundant material for the development of flowers and fruit, began to transpire more freely. But because there was too much liquid, those substances—which are now under investigation and which we shall simply call nectar until they are analyzed—were thinned out considerably, with the result that everything was transpiring at once. However, the dry air immediately removed the watery portions, leaving the more substantial behind on the leaves.

To these the plant lice and other insects now thronged; however, they were not the cause of the phenomenon.

It is somewhat harder to explain just how the honeydew dripped to the earth, evenly splashing some stones and fully coating others. But it would appear to me that when this juice oozed from the leaves, air became imprisoned in the hollows, along the ribs and other parts, the vertical position of the leaves possibly playing a large part. Sun and warmth may then have expanded the air to form a blister, which eventually broke, and in breaking discharged its moisture.

Corroborating this conjecture is the fact that no honeydew could be observed on the flowering lindens, for the preparatory juices which are dissipated in honeydew had already reached their destination and the moisture which in the one case had appeared irregularly had here fulfilled a nobler purpose.

Perhaps the later-blooming lindens absorb less liquid and utilize it more moderately, which would explain why exudation does not occur.

The Reine Claude, however, was just the sort of tree in which to observe in the fruit the great upsurge of juices which it must regularly produce. In such cases, where the fruit was not perfectly developed while at the same time the trunk, limbs, and branches were teeming with nourishment, it was natural to expect excessive exudation. This was not the case with the common plum.

I took advantage of the opportunity to collect a portion of the sticky moisture. Taking about 400 leaves, I tied them in bundles, dipped their tips into tepid water, let them steep for ten minutes, and so on. The moisture dissolved as a piece of sugar does when held in a glass of water and examined against the light, a clear thread twisting itself down to the bottom. The solution in question was now a dirty yellow-green. It was given to Privy Councillor Döbereiner,[53] who analyzed it as follows:

1. uncrystallizable, fermentable sugar,
2. mucous (animal mucilage),
3. a trace of albumen, and
4. a trace of a peculiar acid.

It is to be hoped that the final result of the fermentation to which a part of the honeydew was subjected, will indicate whether manna was also an ingredient. Manna is of course not fermentable.

DÖBEREINER

Jena, July 30, 1820.

With many plants, especially those that are classified as fatty, such exudation occurs with even the earliest organs. *Cacalia articulata* discharges extremely large drops from the young branches and emergent leaves, the stems of which in turn are also destined to become distended members. Among numerous other peculiarities, *Bryophyllum calycinum* exhibits likewise the following one: if younger or older plants are heavily watered, but at the same time have insufficient light and warmth to effect proportionate evaporation, delicate clear drops will emerge from the margins of the leaves, and not from the notches where the young buds will later develop but from the intervening elevations. In young plants the drops disappear after the sun and warmth return; in older plants they coagulate into a gummy substance.

A few more words about evaporation. We find that the pollen, which is charged with the great task of fertilization, may even take the form of vapor. At certain summer temperatures, the pollen vesicles of some pine varieties rise like infinitely small balloons, and indeed in such a mass that when a thunder shower precipitates them they appear to leave a sulphur powder behind.

The seed of *Lycopodium,* readily inflammable, ascends in a flaming vapor.

Other exhalations take the form of sugar, on leaves, branches, stems, and trunks; they may also appear as oil, gum, and resin. If one strikes the proper moment, one may observe the dittany enkindled, a vigorous flame blazing up around stems and branches.

Plant lice, flies, and insects of all kinds find nourishment on certain leaves whose delicate exhalations, except for this indication, would not have been perceptible.

---

[53] Johann Wolfgang Döbereiner, 1780–1849, professor of chemistry, Jena.

On certain leaves raindrops retain their spherical shape and clearness, without dissolving, and this we may undoubtedly attribute to some sort of exudation which, remaining on the leaf, envelops the raindrop and holds it together.

The delicate coating on the skin of a ripe plum is actually clouded and gumlike, but appears blue to our eyes because of the dark background.

It has already been recognized that a certain mysterious reciprocal effect between plants might be beneficial as well as harmful. Is it not possible that certain plants fail to thrive either in cool- or hothouses simply because hostile plants have been given them as neighbors? Perhaps some varieties selfishly take possession of beneficial atmospheric elements intended for all?

Flower lovers maintain that to produce perfect seed, single gillyflowers must be planted between double ones. It would appear, therefore, that the delicate fragrant exhalation might at least increase, if not produce, fructification.

Such influences are assumed to exist even below the earth. It is maintained that bad varieties of potatoes, when laid between better ones, exert a harmful effect. And how many other examples could be cited to persuade or even force those who love this beautiful world, who embrace it with passion, to concede unhesitatingly that all phenomena interpenetrate.

In the development of insects, evaporation is of utmost importance. The perfect though not complete butterfly, emerging from its last caterpillar skin, and now encased in a new skin which foretells its ultimate form, takes along a precious fluid. Organically distilling this fluid, it keeps possession of the more precious part of it while the less important part evaporates slowly or rapidly according to the outer temperature. On closer examination of such natural processes, we have noticed a very marked lessening in weight. Furthermore, when such chrysalises are kept in a cool place, their development is delayed for years, whereas from others which are kept warm and dry the butterfly soon makes its appearance.

We are not saying all this under the illusion that we are saying anything new and remarkable. We are merely pointing out that all things in Nature play and work upon each other, and that developing creatures in their first beginnings, and in their highest forms as well, appear ever the same and ever different.

An Analogous Volatilization. In the fall of 1821 a large caterpillar, evidently that of a copper, was found in a dark spot, just as it was about to fasten itself to a wild rose branch. Placed in a glass with some silk wadding, it used only a few threads to fasten itself more firmly to the glass. A butterfly was now expected to emerge. However, none made its appearance. Instead, after a few months the following remarkable phenomenon was observed. The chrysalis had burst at the bottom and had distributed its eggs over its exterior. But what was more extraordinary, it had thrown them aside individually, against the opposite wall of the glass, a distance of three inches, thus performing an act similar to the volatilization mentioned previously. The eggs, originally full and round, with some suggestion of the grubs they contained, had collapsed and dried up by the beginning of April. Insect lovers are undoubtedly familiar with similar instances.

Increasing Difficulty in Botanical Instruction[54]

I MUST NOT pass over the importance to me of a review of Wenderoth's[55] *Botanical Textbook* appearing in the *Göttingen Review,* No. 22, 1822.

The reviewer, after noting the difficulty of presenting both abstract and concrete botanical information in a textbook, hastens to the main point, which in his opinion may possibly be the source of the reprehensible wavering of almost all recent works on general botany:

"It depends on whether we wish to pursue the plant in its living metamorphosis as a 'something' capable of existence only in regulated alteration, or whether we wish to grasp and retain it as something constant, and therefore dead, in one or several widely separated specific situations. The choice is crucial. Whoever declares himself with Linné for the latter method, takes the safer course. However, once we have ventured into the cycle of metamorphosis, we may no longer hesitate or even turn back.

"Instead, we must pursue the course of development from the first utricle whence fungus and alga, as well as the seed of the highest plant, emerge. We may not derive the higher organs of the plant from the root and stem but solely and singly from the node, from which the root and stem have also developed. And in contemplating the plant as a whole, we must consider it not simply as an individual, but investigate

---

[54] Written in 1821–1822; printed in *Natural Science in General; Morphology in Particular,* Vol. I, No. 4 (1822).

[55] Georg Wilhelm Franz Wenderoth, 1774–1861, professor of botany, Marburg.

how the plant attains its ultimate form by gradually aligning one node after the other, each node having under certain circumstances the power of individual growth. Such study then gives rise almost automatically to a definite genetic concept of species in the plant realm, a concept which many, having set out upon the wrong path, had almost given up. Thus firm ground has once more been gained for a reconsideration of morphology, often defended and often attacked in our times, yet not to be given up by a scientist without risk of sacrificing all accuracy."

Here I should like to append the following thought of my own: an idea cannot be demonstrated empirically, nor can it actually be proved. An individual not in possession of it, will never catch sight of it with his physical eye. The individual who does possess it, easily trains himself to look beyond outer appearances, although returning to reality, after this diastole, to reorient himself. It is possible that he might follow this alternating procedure throughout his life. That it is difficult to deal by this method with didactic or even dogmatic aspects of the subject is no secret for us who understand the value of the method.

As a learned science, botany has a firm footing with regard to professional methodology; as a practical art, it has a firm footing with regard to practical technique. No one has any fear on either score! However, as the abstract keeps intruding, lecturing does inevitably become more difficult. In this we agree completely with our unknown friend and colleague; and no less than he, do we cherish and cultivate the hope he expresses in closing.

Remarkable Healing of a Badly Injured Tree[56]

Around the beginning of the century, the extremely tall and stout mountain ash trees standing of old in the outer court of the home of the Chief Forester, in Ilmenau, began to rot. Orders were given to have them cut down. Unfortunately, the woodcutters started sawing at one that was perfectly sound. They had cut two-thirds through it before they could be stopped; the injured portion was then slatted over, packed, and safeguarded against air. Thus the tree stood another twenty years, until last autumn, when it was broken off at the root during a storm. Its end-branches had already begun to sicken before the storm.

---

[56] The segment of the tree mentioned in the essay was sent to Goethe in December, 1821. The essay itself was written sometime before May, 1822, and published in *Natural Science in General; Morphology in Particular*, Vol. I, No. 4 (1822).

In the center of a twelve-inch-high segment sent me through the courtesy of Chief Forester von Fritsch, the former cut is visible as a deep scar, but as it was fully healed, the storm could not possibly have damaged the healed portion.

It might be said that the tree had been grafted to itself, for care had been taken to protect the damaged place from air as soon as the saw was withdrawn. Thus the life of the extremely thin bark and the sapwood beneath was resumed at once and continuous growth was maintained.

Not so with the wood: once separated it could not again effect a living union. The blocked juices decomposed and the otherwise sound core fell victim to a species of rot.

Remarkable, however, is the fact that the healed sapwood was unable to form fresh wood, the decay of the core, therefore, approaching two thirds.

Not so with the healthy third, for it seems to have continued its growth and thus to have given the trunk diameter an oval form. The shorter diameter, through the middle of the annual ring, measures fifteen inches; the longer diameter eighteen inches, five of which have the appearance of perfectly sound wood.

Problems[57]

NATURAL SYSTEM—a contradiction in terms. Nature has no system; she has, she *is* life and its progress from an unknown center toward an unknowable goal. Scientific research is therefore endless, whether one proceed analytically into minutiae or follow the trail as a whole, in all its breadth and height.

The concept of metamorphosis is a highly estimable gift from above, but at the same time a highly dangerous one. It leads to formlessness, destroys knowledge, disintegrates it. It is like centrifugal force and would lose itself in the infinite if a counterweight were not provided. I am referring to the specification force,[58] that tenacious capacity for persistence inherent in whatever has attained existence, a centripetal force that cannot be disturbed in its deepest nature by anything external. We refer the reader to the genus *Erica*.

---

[57] Written in 1823; printed in *Natural Science in General; Morphology in Particular*, Vol. II, No. 1 (1823). Goethe appended a note explaining that these fragmentary remarks had been jotted down as the result of conversations, his own reflections, and stimulated by the letters of a young friend, Ernst Meyer, lecturer in Göttingen.

[58] The tendency toward specificity.

Since the two forces are simultaneously in operation, we would have to expound them simultaneously in any didactic account. This seems to be an impossibility, however.

Once more, perhaps, we can extricate ourselves only by recourse to art.

We might compare the two forces with tones, which in Nature are ceaselessly reproduced, which in the equal temperament are restricted to the octaves, a process which makes possible a decidedly higher and more effective level of music.

We should need to have recourse to a discussion of art. A symbolism would have to be created. But who is to achieve this? Who is to acknowledge it after it has been done?

In reflecting upon what are called genera in botany, I cannot help feeling, even when accepting them as formulated, that genera cannot always be treated alike.

To be sure, there are genera that do possess a character which they reproduce in each of their species, thus permitting a rational approach. These are not readily dissolved in varieties, and undoubtedly deserve to be treated with respect. I shall merely mention the gentians; any alert botanist will be able to point out a number of others.

On the other hand, there are genera that lack character and permit no assignment of species, inasmuch as they lose themselves in endless varieties. If such families are treated with scientific respect, an end is never reached. Indeed, the scientist himself gets lost, for they elude all classifications, all laws. These genera I have, on occasion, audaciously described as "lascivious," and have ventured to apply the epithet specifically to the rose—though this in no wise detracts from her charm. The hedgerose, *Rosa canina,* might particularly invite such a reproof.

An individual, insofar as he acts significantly, takes the attitude of a lawgiver, first in the field of ethics through recognition of duty, further in religion through professing a special inner conviction of God and godly things, and through restricting himself thereafter to prescribed symbolic ceremonies. The same is true in military and civil law: action and deed are significant only when the individual has prescribed it for himself and others. In the arts, the same thing applies to the subjugation of music by the human intellect; it is an open secret to our times how human intellect, working through the greatest talents, has exerted its influence in the supreme epochs of plastic art. In science, the numerous attempts to schematize and systematize are evidence of the same tendency of the intellect. However, all our efforts must be in the direction of eavesdropping on the methods of Nature herself, so that we may prevent her from becoming obstinate over enforced prescriptions, and yet not be deterred from our purpose through her arbitrary behavior.

An Unjust Demand[59]

MY CRITICS have taken me to task for not considering the root in my treatment of plant metamorphosis. When I first heard this, I found it strange that one should expect me to have done forty years ago what has not been done down to the very present. For the plant root I have as much respect as I have for the Strassburg and Cologne cathedrals. (And of the nature of the latter I am not completely uninformed, for I was happy to be able to turn over to my friend Boisserée[60] a drawing of a part of the cathedral foundation which had been partially excavated in earlier times—as something that might be of interest to him.) But our actual observation of the structure begins at the surface of the earth. The term ground plan is given to that part of the building which has reference to the floor, that part from which the structure then lifts itself upward. The lower part, on which the higher or air-seeking part rests, is left to the skill, judgment, and conscientiousness of the master builder. However, from the excellence and the logic of the structure as a whole, we can easily draw conclusions as to the substructure.

So, too, with the root. I was not concerned with it at all, for what had I to do with an organ which takes the form of strings, ropes, bulbs, and knots, and—thus limited—manifests itself in such unsatisfying alternation, an organ where endless varieties make their appearance and where none advance. And it is advance solely that could attract me, hold me, and sweep me along on my course. Let everyone go his own way. Let him, if he can, look back upon forty years of accomplishment, such as the Good Genius has granted me.

---

[59] These obviously unedited paragraphs, dated June 27, 1824, were found among Goethe's literary effects, and were first published posthumously in the Weimar edition.

[60] Sulpiz Boisserée, 1783–1854, intimate friend of Goethe, and student of Gothic architecture. It was Boisserée who induced Goethe to make a study of the Cologne cathedral, a study which removed his former indifference toward it and Gothic art in general.

Book Reviews

History and Development of the Plant World, by Schelver[61]

OUR OLD FRIEND and fellow student has met to the fullest the hopes and desires which we have clearly outlined in the present issue regarding botany, its foundations, teaching, and transmission to posterity. It may well be that this book is rendered more understandable, more pleasing, and more cogent to me than to others by virtue of the mutual stimulation and enlightenment resulting from my personal acquaintance with the author, begun twenty years ago and since then constantly maintained. At any rate, his great talent has given me the highest pleasure and has strengthened my faith in the benefits of vigorously sustained relations, and the consequent enrichment of both parties.

The prospective reader of the little book is advised to turn first to the third section, on the study of botany, page 78.

He will encounter there the lofty thought that all knowledge as manifested in humanity, all urge to knowledge and activity, should be regarded as a living thing, already containing what it will acquire from without in the course of world history and what it will develop from within.

Here, then, noting and noting down, discovery and observation, experience and reflection, collecting and arranging, classifying and surveying, insight and perspective, content and organization, are ever vitally associated. The first is entitled at the same time to become the last, the lowest the highest, the coarsest the finest; and if centuries, perhaps milleniums, are required for such a climax, then the contemplation of these things becomes all the more worthy and valuable; however, all the more must it be kept free from prejudice. Everything that has been accomplished and achieved, be it ever so insignificant, will then retain its worth. Everything that has been felt and thought will be properly evaluated, and everything will be historically recorded just as it was when it came into existence—side by side or following one upon the other.

In this way we can advance beyond our predecessors without eclipsing them, vie with our contemporaries without affronting them. Indeed, it might not be a mere dream to hope that once this viewpoint were grasped aright, we might all work to each other's advantage. And why should those intellects that forge ahead and soar aloft on eagle pinions, and

---

[61] Written in May, 1822, published in *Natural Science in General; Morphology in Particular,* Vol. I, No. 4 (1822). Other Schelver references, pp. 105, 106, 189, 199, 241.

with eagle eye, not also appreciate the efforts of those that dwell in the dank, low regions of the earth, their eyes equipped to seek the infinite in miniature![62]

An essay by our author in precisely this vein, "The Function of Higher Botany," may be found in the second part of the tenth volume of the new proceedings of the Leopold-Caroline Academy, Bonn, 1821, an excellently designed volume, from which we have already derived great benefit.

Color Chart of Organic Nature,[63] by Wilbrand[64] and Ritgen[65]

(Lithographed by Päringer)

Attempts have been made repeatedly—and undoubtedly will be continued—to present to the eye, by symbolic means, facts which have a sensuous basis yet are not visibly perceptible, so that imagination, memory, and understanding may be stimulated to fill in what is missing. This present attempt has been successful to a high degree.

On a plate 4' 4" long and 1' 10" high, we see first an ocean 8" high. The horizontal line above this ocean extends from 90 degrees north to 90 degrees south. In the middle is the equatorial region, the richest and most diverse in the organic life that extends from here to all sides; the curve of the snow line drops from its highest point in the middle until it finally rests upon the ocean, earlier in the south and later in the north. Above it are the highest snow and ice-covered peaks, an especially imposing view, for the Himalayas are represented here too. And now, by means of lines emerging from a center placed directly on the ocean's surface, fish are listed at the bottom, animals on the right, plants on the left; likewise their climatic distribution is suggested.

Memory and imagination both are stimulated; all information supplied by explorers is symbolically rendered for us; earth and sea are populated. And once the large plate is hung on a wall, one will be unable to do without it, and one will likewise want to have the little book with commentaries constantly at hand.

---

[62] That is, with microscopes.

[63] This work, published in Giessen, 1821, was dedicated to Goethe, Alexander von Humboldt, and Johann Friedrich Blumenbach (see p. 233 for another reference). This review was written in 1822 and printed the same year in *Natural Science in General; Morphology in Particular,* Vol. I, No. 4 (1822).

[64] Johann Bernhard Wilbrand, 1779–1846, professor of anatomy, Giessen. See p. 241.

[65] Ferdinand August von Ritgen, 1787–1867, Privy Medical Councillor in Giessen.

The two men must be felicitated on having discovered each other and on having co-operated in their work, and then on finding a technician skilled enough to execute the plates faultlessly.

We must also praise the splendid coloring and point out that it is helpful in grasping the concept. For that reason it is to be hoped that all copies are as beautifully illuminated as the one we have before us.

In our last issue[66] we briefly and approvingly took note of Wilbrand and Ritgen's color chart. Since that time it has had a place on the wall of our study, and we can repeat our recommendation that it is most helpful in all scientific observations of earth and Nature. A number of visitors, inspired by the imposing sight, have expressed the desire to own a smaller copy.

However, the chart can be procured commercially only without illumination, but since color alone can properly indicate the characteristic relationships, such an uncolored chart is unsatisfactory. I present this matter to our worthy authors for consideration, and inquire of them whether it would be possible for them to have orders for illuminated copies taken by subscription,[67] in view of the fact that accurate coloring of the charts through other channels might be harmful to the publishers. We hope that our good will and grateful interest is apparent.

During our study of this succinct symbolic chart a new, carefully executed work has come to our assistance, one which we like to keep close at hand.[68]

Nature, Its System and History, by F. S. Voigt[69] (1823)

For many years we have been witness to the author's tireless activity in Nature's splendid realm, and it is a great pleasure now to see the imposing results of his studies and reflections in book form. At some later time we shall discuss in detail the benefit we derive from it. Already it has given us considerable assistance: the organization of the book as a whole has stimulated us to thought; we have used it for short, concise references; and then, when we felt the need of proceeding further, we were aided by its helpful and accurate bibliography. From this informative work many a student of natural science will undoubtedly

---

[66] This additional note appeared in *Natural Science in General; Morphology in Particular,* Vol. II, No. 1 (1823).

[67] Goethe's suggestion was later adopted.

[68] Goethe is referring to the work reviewed immediately following.

[69] This notice was printed in the same issue as the second notice of the Wilbrand-Ritgen chart. Friedrich Sigmund Voigt, 1781–1850, botanist, Jena. For other references to Voigt, see pp. 189, 198, 201, 206.

derive whatever he needs by way of instruction, criticism, assistance, suggestions, and all the other advantages expected of a textbook. And we do not doubt that a good many others have already had experiences similar to our own, and likewise cherish a grateful feeling toward the worthy author.

Notes for a Physiognomy of Plants,[70] by Alexander von Humboldt
Read at an open session of the Prussian Academy of Sciences, Berlin, January 30, 1806

Our first ardent wish was fulfilled in hearing that this outstanding and intrepid scientist was safe at home again after his arduous and dangerous travels; but a second wish, equally acute, followed immediately upon the first, for everyone has been eager to hear an account of the abundant treasures he has acquired. Here we have the first gift, a basket of most precious fruit.

When we venture into knowledge and science, we do so only to return better equipped for living; and here, indeed, the science of botany, still so pitifully backward in individual details, appears upon a height, transfigured, affording us strong and singular gratification.

Linné created an alphabet of plant forms and provided a convenient list; Jussieu arranged the entire field into a system more harmonious with Nature; sharp-sighted men, with and without microscopes, continue to classify distinguishing features with extreme accuracy and down to the smallest detail; and philosophy is promising to furnish a vitalizing spark in a unified and higher point of view. Now Humboldt, with access to plant forms in living groups and masses throughout the world, takes the final step in advance of all others, indicating the way in which facts—individually recognized, comprehended, and studied—can be appropriated by the intellect in full splendor and opulence, and how the firewood, long stacked and smoldering, can be fanned through aesthetic inspiration into a bright flame.[71]

---

[70] Goethe received Humboldt's lecture on February 22, 1806, and this review of it appeared in *Jenaische allgemeine Literatur-Zeitung* (Jena General Literary Journal) within the year.

[71] There follows a long passage from Humboldt's lecture, of no concern to us here.

Genera et species palmarum, by Martius[72]

Travel Book of a Scientific Expedition to Brazil
Physiognomy of Plants

Genera et species palmarum by Dr. C. F. von Martius, Pts. I and II, Munich, 1823. The two brochures contain forty-nine lithographed plates illustrating various species of native palms observed by the author on a scientific expedition to Brazil several years ago.

The plates representing details of branches, leaves, flowers, and fruit resemble delicately etched copper plates, skillfully executed. In this respect they may undoubtedly be compared with the beautiful osteological copper engravings in Albinus' work, and may possibly even surpass them in delicacy of execution. Most of them were done by A. Falger; but those of J. Päringer and L. Emmert are likewise of outstanding merit.

Ten pages of charcoal drawings, neatly and vigorously drawn in the usual manner, illustrate various species of palm trees, full length with trunks and branches, suitably accompanied by a view of the regions in Brazil where these palm varieties thrive best. In addition, the richly detailed foregrounds acquaint the reader more fully with other plants and with the extreme lushness of the country's vegetation. To amplify what we have just said, we need only suggest in general what each plate represents.

Plate 22. Chief subject: *Oenocarpus distichus;* in the foreground, leaves and shrubs, in the mid- and background, low-lying meadows between wooded hills.

Plate 24. *Astrocaryum acaule* and *Oenocarpus Batava,* the chief subjects, appear in the foreground; the surrounding landscape is the low shore of a gently flowing stream, flanked on both sides by richly wooded promontories.

Plate 28. *Euterpe oleracea,* likewise on the shore of a river flowing into the sea, whence the floodtide is pouring in.

Plate 33. Striking the eye first of all are the *Elaeis melanococca* and *Iriartea exorrhiza,* then a wooded mid-ground and low bank of a river or lake; a crocodile just emerging from the water enlivens the landscape.

Plate 35. *Iriartea ventricosa,* together with a view into a narrow

---

[72] Review by Goethe written and published in 1824 in *Natural Science in General; Morphology in Particular,* Vol. II, No. 2 (1824). Carl Friedrich Philipp von Martius, 1794–1868, professor, Munich. The scientific expedition mentioned in the title was headed by Martius and Johann Baptist Spix, 1781–1826, zoologist and director of the museum of Munich.

gorge formed by wooded mountains growing higher and higher, from which a river pours forth, creating a small waterfall in the foreground.

Plate 38. First *Mauritia vinifera;* in the background, desolate hills; the plain is sparsely covered with this variety of palm.

Plate 41. *Attalea compta* and *Mauritia armata;* in the background, an almost desolate region, with only a few additional trees of this species in the near and far distance.

Plate 44. In the foreground, *Mauritia aculeata;* in the background, an impenetrable thicket of trees and also large-leaved, treelike plants.

Plate 45. *Lepidocaryum gracile* and *Sagus taedigera* in a dark forest region permitting no distant vista.

Plate 49. *Corypha cerifera.* The landscape which serves as background represents a plain richly overgrown with trees, especially palms; in the distance, towering mountain tops.

The useful and instructive content of these plates will undoubtedly be clear, in spite of the brevity of the foregoing descriptions. But perhaps we should note, in addition, that the taste and artistic manner with which Herr von Martius has arranged the subjects into a landscaped whole, is deserving of praise on the part of those who can judge the work from the standpoint of art. Experts will be no less satisfied with the work of Herr Hohe,[73] who drew the last-mentioned illustrations on stone plates, in the manner of charcoal drawings, from original sketches by Herr von Martius.

This work, meritorious in so many ways, has been viewed solely from the artistic-aesthetic aspect in the foregoing passages. However, we may say that it is precisely this artistic quality which should be regarded as a correlative advantage of the expedition of these excellent men.

The well-known travel book which Herr von Spix and Herr von Martius published in 1823 in Munich, gave us extremely welcome local scenes covering a wide area, grandiose, unconfined, and extensive. It gave us most diversified information on individual phases, stimulating to both the imagination and memory. But what gave this spirited presentation its particular charm was its appreciation of the sublimity of Nature in all her aspects, its pure, warm, devoutly profound appreciation—experienced and expressed with clarity and enthusiasm.

In addition, their book *Plant Physiognomy,* Munich, 1824, enables us to view from a high vantage point an otherwise unsurveyable portion of the earth's surface. It calls attention to extraordinary features, to climatic and local conditions amid which the innumerable plants thrive

---

[73] Friedrich Hohe, lithographer and painter.

and group themselves, and it presents all this in such abundance that only an experienced botanist would have sufficient terminology at his command to assign names to all the subordinate forms.

In this latest book, which we have given more detailed study, it is likewise by means of an elaborate technical language that the rare species of the palm family are so richly described for the professional scholar. However, consideration is also given to the amateur, for he finds in the plates the chief characteristics and forms of the universal primitive state, all represented in great variety. He sees isolated or grouped settlements and dwellings, situated on moist or dry, high or low land, in open or wooded locations. Thus knowledge, imagination, and feeling are all stimulated and satisfied. As we peruse these publications we feel thoroughly at home in this remote continent, much as though we had been there ourselves.

Bignonia Radicans[74]

IN SEPTEMBER of 1786, in the Botanical Garden of Padua, I gazed upon a broad, high wall completely covered with *Bignonia radicans*. The clusters of deep-yellow, chalice-like flowers, growing in endless luxuriance, made such an impression upon me that I became especially attached to this plant, always devoting special attention to it when I encountered it in botanical gardens, in my own garden, and in the parks of Weimar, where it was a favorite.

It is a twining plant, and seems to have an inclination to propagate itself into infinity; however, it lacks the organs for nestling, clinging, and adhering. By tying it to a trellis we are able to keep it upright and force it to climb very high.

This method, which I have observed from time untold, I continued in my own plantings, but noted almost with annoyance that the new branchlets withdrew behind the trellis toward the wall. As they pressed against the wall, they sometimes awkwardly squeezed the lovely flower clusters, thus hiding them from the sight of spectators desirous of admiring them. After all sorts of observations and investigations I finally discovered the following:

If I take a branch of *Bignonia*, I can see that odd-pinnate leaves emerge from it; that below them, on the reverse side, glandlike excres-

---

[74] Written in August and September of 1828, and first published posthumously in the Weimar edition.

cences appear, which look like grapes when observed under the magnifying glass. The three middle descending series have approximately fifteen of the berries or beadlets; the following ones have fewer; and in this way the effect of a grape cluster is brought about. As mentioned, two such organs stand side by side beneath the leaf-pair at the rear end of each node. The beadlets of these grapelike clusters are clear and beautiful in their beginnings; however, I myself have observed them only once—at the end of August, 1828—but I shall be on the alert for this phenomenon next spring. Meanwhile, I have put my present specimens into alcohol, where their form is fully preserved, though their color has turned brown.

Moreover, this beadlet, about one unit high, often has the appearance of cork, being brownish and dry, resembling a bristly cockscomb. One might consider it to be a worthless and perhaps harmful excrescence. In shape, it does not generally remain the same but extends its length down along the stem, isolates itself in little clusters, is lost in little depressions, which give way to little grooves extending down to the wood. Only once did I find a really strong cluster, nine units in height, branching out like genuine roots; when observed through the microscope, its delicate fibres proved to be covered with fine hairs.

The question is, then, whether these places would not actually strike root under the proper conditions; at least, one cannot but regard these organs as moisture conduits, very much needed by vines which are several years old and which are therefore far removed from the root and the soil.

In very young branches of *Bignonia* trained high against a wall, we find no trace of such an organ; but on a plant unfavorably located on a moist spot with little sun, and grown to a bush hardly an ell in height, we find the branches set with these organs at several nodes. Thus the reciprocal effect becomes clear: the organ is produced by the very moisture which it is to impart to the plant.

I tell myself therefore: this is a twining, but not a climbing plant, and we mistreat it when we force it to go upward, where it cannot get proper nourishment. Taken upon a height and from there allowed to play down upon terraces and rocks, it will display itself in greatest beauty. The under side of the youngest branches can then rest upon the damp stone and absorb sufficient moisture to produce unusual greenness and thousands of flower clusters. In that way the branchlets, too, are in a natural position. For in the usual practice, a branch leaning upward against the wall is bowed over by the weight of flowers, causing the under side, which is just developing these organs of nourishment, to be turned to the light and the sun. Thus, just at a moment when the development of the plant requires such influences, the live organs are dried and destroyed. Likewise the leaves of the branch drop off, leaving the flower cluster with a denuded stalk, whereas it might have been covered with leaves down to the very flowers.

We may plant the grapevine to twist and trail wherever we deem suitable, for it can fasten itself anywhere by means of its forks. But a strikingly beautiful plant like *Bignonia radicans* should be planted high and allowed to grow downward. Located in this manner in a sunny spot, it will soon put forth a wealth of golden clusters, whereas until now this strikingly decorative plant, though cultivated with special care, has yielded only limited results.

For the benefit of those wishing to write monographs on this plant, I must still mention, in closing, that on the individual leaf stalks of the pinnate leaves, just below the leaf-beginning, six to eight of such aforementioned glands may be found; also that on the branches, at the point where one node gives way to another, just below or beside the buds, a series of extremely delicate hairs emerge; and finally, that the intervals of the branch from node to node are set with numberless white points in such a way that no part of the plant is deprived of the means of absorbing from the atmosphere or from its environs the moist nourishment it requires.—*Dornburg, August 26, 1828*

The Spiral Tendency[75]

In several scientific lectures at gatherings of German scientists in Munich and Berlin,[76] our learned and brilliant Martius[77] succeeded in rounding out all the evidence acquired to date in support of the theory of morphology, by pointing out the tendency in plants which actually shapes and determines blossom and fructification,[78] and which we should like to call the spiral tendency. As given in the 1828 and 1829 issues

---

[75] Goethe's first treatise on the subject, published in the last year of his life, in J. W. von Goethe, *Versuch über die Metamorphose der Pflanzen,* Uebersezt von Friedrich Soret, nebst geschichtlichen Nachträgen (Stuttgart: in der Cotta'schen Buchhandlung, 1831), [with parallel title]: J. W. de Goethe, *Essai sur la Métamorphose des plantes,* traduit par Frédéric Soret, et suivi de notes historiques (Stuttgart: J. G. Cotta, Libraire, 1831).

[76] 1827, in Munich; 1828, in Berlin.

[77] See pp. 17, 123, 124, 136.

[78] Martius' lecture was entitled "Architectonics of Flowers."

of *Isis*,[79] he states his views as follows: "This advance in the knowledge of plant life is the result of that morphological view[80] known as metamorphosis.

"All organs of the flower—calyx, corolla, stamens, and fruit buds—are transformed leaves. Hence they are leaves that are essentially alike, differing only in their morphological potentialities.

"Accordingly, the construction of a flower depends on the position and arrangement, peculiar to each species, of a certain number of metamorphosed leaves. These leaves, inwardly similar but outwardly diverse, arrange themselves around a common axis near the end of a branch or sometimes peduncle, until, through union and mutual connection, they come to a standstill."

Up to this point, we have given only the most essential ideas in the author's own words, arranged, we hope, in a way our worthy author would approve. We merely add the following:

This masterly scientist then treats these numerically and dimensionally ordered organic movements of what is inwardly alike, and yet outwardly completely different, in such a way that he is justified in giving them the name of "organic rotations." Also, by means of analyses of all kinds, he penetrates so close to both the regular and irregular phenomena that he can venture to use a symbolic terminology for the details and to base a new system upon it.

A study of the essays mentioned, a private conference with the worthy man himself,[81] and an apparatus designed to convey a clear idea of this problematic natural phenomenon, enabled me to pursue these important views and to acquire an opinion that I shall not hesitate to present here after inserting the following passages for fuller understanding.

The spiral vessels are fairly well known to botanists in general, especially to anatomical botanists. The vessels have been observed in all their diversity, sorted and named, even though their actual classification is still considered problematical. Here, however, we shall regard them as extremely small parts identical with the whole of which they are a part, which should be regarded as homoeomeriae,[82] transmitting their char-

---

[79] Publication of Lorenz Oken, 1779–1851, the initiator of these conventions of German natural scientists and physicians.

[80] Goethe's.

[81] In October, 1828, when Martius passed through Weimar on his way home from a gathering of scientists in Berlin.

[82] According to Anaxagoras, the properties of a thing are determined by those of its component parts ("seeds"). Later Greek philosophers gave them the name *homoeomeriae* ("similar particles"). See pp. 24, 140.

acteristics to the whole and receiving characteristics and tendencies from it in turn. Independent life is attributed to them, also the power to move independently and to assume a definite direction. The brilliant scientist Dutrochet calls this a *vital incurvation.*

Turning aside from the discussion of these constituent parts, we shall again pursue the regular course of our demonstration.

We had to assume in vegetation a general spiral tendency, by means of which, in combination with a vertical force, all plant structures, all plant formations, are completed according to the law of metamorphosis.

The two chief tendencies then, or, if one will, the two vitalizing systems whereby plant life is perfected through growth, are the vertical and spiral systems. Neither can be imagined apart from the other, for the vitality of one is maintained only through the operation of the other.

But for more exact evaluation, especially in a demonstration, it is necessary to investigate them separately, since the characteristics of this inseparable pair become all the clearer as we observe how one or the other has the supremacy—sometimes overpowering its adversary and sometimes being itself overpowered—or merely manages to maintain a position of equality.

The vertical tendency is manifest from the very beginning of germination onward, for it is this tendency that enables the plant to take root and at the same time to lift itself upward. It persists from beginning to end, revealing also a solidifying function, either in the elongated fibres and threads or in the straight and rigidly erect structure of the wood. It is also this same natural power that ceaselessly forces itself upward, or in some other direction, from node to node, carrying individual spiral vessels along with it. Thus by fostering and increasing life upon life, it consistently produces a continuity of the whole, even in climbing or creeping plants.

However, it is seen most clearly in the inflorescence, in that it forms the axis of every flower formation. And it is most striking in the spadix, in the spathe, where it clearly emerges as staff and support of the final fulfillment. For this reason the vertical tendency must be kept continually in mind even in studying newer viewpoints, and must be regarded as the virile sustaining principle of growth.

The spiral tendency, on the other hand, must be regarded as the actual reproductive life principle. It is closely related to the vertical tendency but is usually relegated to the periphery of plant growth. At the same time, it can occur as early as the first germination, as we can clearly note in several bindweeds.

It is most strikingly exhibited in terminations and conclusions. Take, for example, the way so-called composite leaves often end as tendrils and vrilles; also the way whole branchlets, in which the juicy veins attain supremacy and in which solidification is lacking, gradually or rapidly assume tendril forms such as forks, caprioles, and the like.

In the case of monocotyledons, the spiral tendency is less often evident during the course of growth. Here the vertical or longitudinal tendency appears to predominate: leaves and stems are forced up by erect fibres. For example I have encountered neither cirrus nor vrille in this large category of plants.

But the spiral tendency, whether it conceals itself or emerges perceptibly during the development of the plants, at the end, during blossoming and fruiting, predominates. Then, in thousandfold twistings around its center, it performs the miracle that enables a single plant to derive infinite reproduction from within itself.

Herewith we again return to our beginning, to recall the words that originally led us to such multiplicity of thought.

If the foregoing provides the desired explanation for ordinary plant formation, further reflection and investigation will fully reveal that these same principles will provide a basis for judging the extremely diverse misgrowths that appear as deviations from the law of definite forms.

More detailed investigation is needed for both more profound and more exact information. This we have good hope of attaining, for Martius himself cannot fail to continue this important subject, and younger men also are making vigorous and thoroughgoing efforts to work out the perceptible and calculable limits of the rotations. Thus, for the present we need make but general and admiring reference to an article appearing in the first part of Volume XV of the proceedings of the Leopold-Caroline Society.

The treatise is entitled "Comparative study of the classification of the scales of fir-cones, as an introduction to the study of leaf placement in general," by Dr. Alexander Braun.[83]

As for us, we need merely add the wish that in the study of this subject, once more being carried on in such infinite detail, there may be no want of integration, so that the general perspective growing out of such rich experience may be captured and preserved within the limits of an exact science.

---

[83] Alexander Braun, 1805–1877, German botanist, for many years director of the Berlin Botanical Garden.

On the Spiral Tendency in Plants[84]

Preliminary Notes, Aphorisms

WHEN A CASE occurs in our study of Nature which takes us by surprise and in which our customary concepts and modes of thought prove inadequate for mastery of the subject, we do well to cast about to see whether something similar has already been treated in the history of thought and understanding.

In the present case, we are reminded of Anaxagoras' *homoeomeriae*. To be sure, a man of his day had to be content with explaining the same by the same. We ourselves, supported by experimental evidence, may be more venturesome in our interpretations.

Let us disregard the fact that these homoeomeriae are more easily applied to primordial, simple phenomena.[85] However, we have discovered here, on a higher level,[86] that spiral organs pervade the whole plant down to its smallest part, and we are likewise sure of the existence of a spiral tendency whereby the plant completes its life course and finally attains completion and perfection.

Therefore, let us not cast aside completely that idea of Anaxagoras. Let us remember that the opinion of an outstanding man is always of value, even though we may not understand immediately how to master and apply what he has said.

With this newly opened vista, we now venture the following statement: after one has completely understood the metamorphosis of plants, one must next consider the *vertical* tendency to become more familiar with the development of plants. The vertical tendency should be looked upon symbolically as a staff, basic to existence and capable of preserving it for a long period. This vital principle is manifested in the longitudinal fibres used as pliable threads for various purposes; it is what creates the wood of trees, keeps the annuals and biennials erect, and even brings about the extension from node to node in vines and creeping plants.

Next, however, we must observe the spiral tendency entwining itself around the vertical.

The vertically ascending system in plant formation produces the

---

[84] This treatment of the subject, more comprehensive in its plan yet more fragmentary in execution than the foregoing essay, was found among Goethe's literary effects after his death and first published in 1833 in the posthumous works (see "Bibliographical Note," p. 255). The two fragments together indicate the general outline of Goethe's views on the spiral tendency.

[85] That is, to purely physiochemical phenomena of inorganic life.

[86] That is, with higher plants.

durable, eventually solidifying, and permanent parts, namely, the fibres of temporary plants, and the greatest portion of the wood in enduring plants.

The spiral system is the developmental, reproductory, and nourishing element. As such, it is temporary and almost independent of the vertical; operating in excess, it is soon exposed to ruin, and perishes; joining the vertical, it fuses with it to form a lasting union as wood or some other solid.

Neither of the two systems can be imagined alone; they are ever and eternally one; and in complete equilibrium they produce the most perfect vegetation.

Since the spiral system is the nourishing system, bud after bud developing in its course, it is self-evident that an excessive supply of nourishment will give it the ascendancy over the vertical, and that the whole complex, as though deprived of its skeletal structure, will hastily develop an excess of buds and disappear.

For example, I have never found the flattened curved ash branches, called bishop's crooks[87] in cases of extreme abnormality, on high, full-grown trees, but only on truncated ones, where excessive nourishment is supplied to the new branches by the old trunk.

Other abnormalities, later to be presented in greater detail, arise when that upward-striving force is thrown out of equilibrium with the spiral and is outdistanced by the latter. In such cases the vertical construction is weakened, and in plants which produce either fibres or wood, it is thwarted and almost destroyed; on the other hand, the spiral system, upon which the embryos and buds depend, is accelerated; the branches are flattened; and the plant stem, which is lacking in wood, is distended and its interior destroyed. All during this process the spiral tendency makes its appearance, expressing itself in windings, crooks, and twists. By studying these examples, one will obtain a basic text from which to draw conclusions.

The spiral vessels, the existence of which has long been known and fully recognized, must therefore be regarded only as individual organs subordinated to the whole spiral tendency. They have been sought and found almost throughout the whole plant body, especially in the splint, where they even emit signs of life. Indeed, nothing is more typical of Nature than that she carry out her whole intention into the smallest division.

This spiral tendency, as the fundamental life principle, must appear

---

[87] See pp. 100–101.

first of all in the development of the plant from the seed. Let us now observe it as manifested in dicotyledons, where the first seed leaves appear definitely paired. It is true that in these plants a second pairlet of more developed leaves is already lodged above the crux and that this sequence persists for a time. Nevertheless, it is obvious that with many other plants the upward-following stem leaves and the bud, actually or potentially lodged behind it, do not always get along together and are always trying to hurry ahead of each other. From this situation the most wonderful positions result, and finally, through rapid converging of all parts of such a series, the approach to fructification in the flower inevitably follows, and eventually the development of the fruit.

In the calla the leaf ribs very soon develop into leaf stalks, growing round gradually, until they finally appear entirely rounded as a peduncle. The flower is obviously a leaf-end which has entirely lost its green color. Its vessels, which proceed without branching from the leaf-beginning to its periphery, are entwined around the spadix inwardly from the outside, and the spadix now maintains the vertical position in the inflorescence and fructification.

The vertical tendency is revealed from the first beginnings of germination onward. It is by means of this tendency that the plant takes root in the earth and simultaneously lifts itself upward. We must indeed consider the extent to which the vertical tendency exercises its rights in the course of growth, inasmuch as we ascribe to it sole responsibility for the right-angled alternate position of dicotyledon leaf pairs. On the other hand, this interpretation might seem doubtful, since a certain spiral influence during the upward climb cannot be denied. At all events, no matter how much the spiral tendency may have withdrawn, it will appear again in the inflorescence, since it forms the axis of every flower formation, manifesting itself most clearly in the spadix and in the spathe.

Anatomical investigations have at length cleared up the subject of the spiral vessels, which penetrate all vegetable organisms, and their deviations in form also. We shall not treat this as a subject by itself, since the elementary student can find the material summarized for him in compendiums, and the progressive expert can search it out for himself in leading works and through personal observation of Nature.

It has long been suspected that these vessels give life to the plant, but the process itself has not been sufficiently clarified.

Recently serious attempts have been made to recognize and present them as being in themselves alive, as the following essay proves.

Edinburgh New Philosophical Journal
October–December, 1828, page 21

"On the universal presence of the spiral vessels in the plant structure," by David Don.[88]

"It has been generally believed that the spiral vessels are seldom found in the parts of the fructification; however, repeated observations have convinced me that one encounters them in almost every part of the plant structure. I have found them in the calyx, corolla, stamens, and style of *Scabiosa atro-purpurea* and *Phlox,* in the calyx and coronal leaves of *Geranium sanguineum,* in the perianthium of *Sisyrinchium striatum,* in the capsules and petiole of *Nigella hispanica;* and they are also present in the pericarp of the *Onagraceae,* composites, and *Malvaceae.*

"In these observations I have been guided by the ingenious observations of Mr. Lindley,[89] who reports in the last issue of the *Botanical Register* on the structure of the seeds of *Collomia,* which he represents as being enveloped by an intertwining of spiral vessels. These vessels in the *Polemoniaceae* seem to be analogous with the hair or pappus with which the seeds of certain types of the *Bignoniaceae, Apocynaceae,* and *Malvaceae* are provided. But further observations would be necessary before concluding that they are genuine spiral vessels. Spiral vessels occur very frequently in the stems of *Urtica nivea, Centaurea atro-purpurea, Heliopsis laevis, Helianthus altissimus, Aster Novi Belgii,* and *A. salicifolius,* in all of which they are visible to the naked eye, and are therefore recommended to botanical amateurs as obvious examples of spiral vessels. The stems, gently split longitudinally and held apart at the upper end with a little wedge, show these vessels much more clearly than in cross section. Sometimes these vessels are found to be situated in the pith in *Malope trifida* as well as in *Heliopsis laevis;* but their origin can easily be traced between the wood fibres. No trace of the vessels has been found in the outer bark, but they are found in the splint of the inner rind of *Pinus* as well as in the albumen. However, I have never succeeded in finding them in the leaves of this genus, nor in *Podocarpus,* and they seem to occur more rarely in the leaves of evergreens. The stems and leaves of the *Polemoniaceae,* irises, and mallows are likewise frequently provided with spiral vessels, but nowhere do they occur so frequently as in composites. They rarely occur in the *Cruciferae, Leguminosae,* and *Gentianaceae.*

"I have often noticed that the spiral vessels move vigorously when I

---

[88] David Don, 1800–1841, British botanist.
[89] John Lindley, 1799–1865, professor of botany, London.

detach them from the young powerful shoots of herbaceous plants. This movement lasts several seconds and appears to be not mere mechanical action but the operation of the life principle itself, similar to its operation in animals.

"While I was holding between my fingers a small section of the rind of *Urtica nivea* which I had just cut from the living stem, my attention was drawn for a moment to a marked spiral-like movement. I repeated the experiment with other parts of the rind, and in each case the movement was similar to the first. It was obviously the operation of a contracting force in the living fibre, for the movement ceased after I had held the section in my hand several minutes. It is to be hoped that my short report will direct the attention of scientists to this odd phenomenon."

Bulletin des sciences naturelles
Nro. 2, Février 1829, p. 242

Lupinus polyphyllus. A new species found by Mr. Douglas in the American northwest. It is herbaceous, vigorous, similar to *Lupinus perennis et Nootkatensis,* though larger in all dimensions; and the stem leaves, 11 to 15 in number, are lance-shaped; moreover, there is some difference in the formation of the calyx and corolla.

Mr. Lindley noticed that the inflorescence of this plant provides a significant example favoring the following theory, namely, that all organs of a plant actually stand alternately in a spiral direction around the stem, their common axis; and that this situation holds true even in instances where it may not occur exactly as here described.

"Recherches anatomiques et physiologiques sur la structure
intime des animaux et des végétaux et sur leur
mobilité," par M. H. Dutrochet,[90] 1824.

(*Revue française*, 1830, Nro. 16, pag. 100 sq.)

"The author has directed his experiment principally to 'sensitives,' which exemplify in the highest degree the phenomena of irritability and mobility in plants. The actual principle of movement in this plant rests in the enlargement located at the base of the leaf stalk and at the points where individual leaf parts are dovetailed by means of pinnules. This little swelling is formed by the development of the rind parenchym and contains a great quantity of spherical cells, the walls of which are covered

---

[90] See pp. 129, 200.

with nerve bodies. These are also quite numerous in stem leaves, and later they are frequently found in the sap that runs out when a young branch is cut off from a sensitive.

"The development of the rind parenchym, which has the most significant share in the swelling on sensitives, surrounds a center formed by a bundle of reeds. It seemed important to find which of the two parts represented the actual organ of movement. When the parenchym was removed, the plant continued to live, but had lost the power of movement. This experiment shows, therefore, that the power of movement has its location in the rind part of the distension and that it can be compared, at least in its function, to the muscular system of animals.

"Monsieur Dutrochet has, moreover, discovered that little sections, when cut off and thrown into water, will move and describe a curved path, the deep side directed in every case to the middle of the swelling. He applies to this movement the general term 'incurvation,' and regards it as the element of all movement occurring in plant life, and even in animals. The incurvation is manifested in two ways; the author calls the first an oscillating incurvation, because an alternation of repulsion and attraction can be noted; he calls the second a fixed incurvation, because no such alternation of movements exists. The oscillating incurvation is observable in the sensitives; the fixed is seen in the vrilles and sinuous stems of the convolvulus, clematis, bean, etc. Monsieur Dutrochet concludes from all these observations that the irritability of sensitives has its origin in a vital incurvation."

The foregoing exposition, which greatly clarifies the subject, came to my knowledge quite late, after I had already taken a lively interest in the wider views of our worthy Martius.[91] In two successive annual lectures in Munich and Berlin he had given a clear and comprehensive account of his views. During a friendly visit paid to me on his way home from Berlin, he discussed this difficult subject and drew some hasty but illuminating sketches for me. It was now easier to understand the essays in *Isis,* 1828 and 1829, and, through the kindness of Martius, I was able to duplicate the model described there. This model has rendered good service by graphically showing how calyx, corolla, and the organs of fructification originate.

In this way the important subject was worked out and applied in a practical and instructive manner; and once a progressive man like Martius

---

[91] See pp. 17, 123, 124, 127.

has turned his attention to the most elementary plants, the acotyledons,[92] to discover the beginnings of this universal tendency, we can expect that in due course the theory will be fully worked out.

In the meantime, I am permitting myself to travel the middle road. In my own way, through general observation, I shall try to see whether the beginning may be connected with the end, the first with the last, the well-known with the new, the definite with the doubtful. In this attempt I should like to enlist the aid of my fellow scientists, since my work is not intended to be definitive but merely to serve as a contribution.

We had to assume in vegetation the existence of a general spiral tendency which, in combination with the vertical tendency, effects all plant structure, all plant formation in accordance with the law of metamorphosis.

Thus the two chief tendencies—or if one will, the two vital systems—by which growth is achieved in plant life, are the vertical and the spiral systems. One system cannot be imagined apart from the other, for only the two working together can achieve a vital effect. However, for more precise insight and especially for clearer demonstration, it is necessary to discuss them separately. It is also necessary to see which is predominant in a given instance, for one is constantly overpowering the other or being itself overpowered; or else an equilibrium is brought about. Through careful observation the properties of this inseparable pair cannot fail to become more graphic.

The vertical system, mighty but simple, is the one whereby the upper visible part of the plant is separated from the root and lifted upward in a straight direction; it is predominant in monocotyledons, the leaves of which are indeed formed from straight fibres that can easily be separated from one another and used for various purposes. Here we might mention *Phormium tenax*;[93] the leaves of palms also consist of straight fibres throughout, connected during the earliest period of the leaves but later separated and diversified through continued growth in accord with the laws of metamorphosis.

In the case of monocotyledons, the stems often develop directly from the leaves, which first distend and then develop into hollow reeds. Next, the spiral tendency manifests itself in the arrangement of the three leaf-tips around an axis at the tip of the stem, and from this position the flower and fruit cluster arise, as in the case of the leek family.

The vertical tendency is, nevertheless, noticeable beyond the flower, taking possession of the inflorescence and fructification likewise. The

---

[92] That is, the cryptogams.
[93] A New Zealand flax.

straight-growing stem of *Calla aethiopica* exhibits both its leaf nature and the spiral tendency, when its one-leaved flower, winding about the top, is vertically pierced by the pillar bearing the blossom and fruit. Further investigations ought to be made with arum, maize, and other plants, to see whether the fruits follow each other spirally around this pillar.

In all events, this columnar tendency must be studied with care at the climax of the growing process.

In our investigation of dicotyledons we encounter a conflict between the vertical tendency, whereby the successive development of stem leaves and buds is fostered in sequence, and the spiral system, whereby the fructification is to be completed. A perfoliate rose provides a splendid example.[94]

On the other hand, we have in this very class the most decisive examples of a triumphant vertical tendency and elimination as far as possible of the opposite tendency. We shall mention only the common flax, of general usefulness because of its decided vertical formation. The outer hull and the inner filament rise sharply upward in close unison, and much effort is necessary to separate the chaff from the fibre. How indestructible and strong the thread is, we see by the great reluctance it shows in relinquishing the connection given to it by Nature. Through chance the retting of a plant once resulted from a winter's sojourn under the snow, but the thread merely became stronger and more beautiful in the process.

But after all, there is no need to cite additional evidence, for throughout our lives we see around us linen which by repeated washing and bleaching attains and reattains the dazzling white, the elementary appearance of pure earthy materials.

At this point, where I intend to leave my study of the vertical tendency and turn to the spiral, the question arises whether the alternate position of the leaves on the growing stem of dicotyledons is an inherent part of the spiral or the vertical system. To me, I must confess, it seems that it should be ascribed to the vertical system, that precisely through this type of production the upward striving is carried out in a vertical direction. To be sure, this position can be taken by the spiral tendency, in a certain sequence and under set conditions and influences, but during this process the vertical tendency appears unstable, ultimately becomes imperceptible, and even disappears completely.

---

[94] See p. 73.

But now we reach the point at which the spiral tendency clearly emerges.

We previously declined to discuss the much-observed spiral vessels, though we did give them their due as homoeomeriae, i.e., parts representing and constituting the whole. Nevertheless, we shall not neglect the simplest plants, known as *Oscillaria,* now visible to us with the aid of the microscope. They prove to be spiral in shape throughout; and their existence and growth exhibit movements so odd that one wonders whether they ought not to be counted among the animals. Only a more extensive study and deeper insight into Nature will give us clearer understanding of universal, boundless, and indestructible life. For that reason, we shall gladly believe the statement of the observer mentioned above,[95] that the fresh rind of a nettle manifested a decided spiral movement.

Returning now to the actual spiral tendency, we again refer to the previously mentioned work of Martius, which presents this tendency as reaching the acme of its power in the climax of flowering. We shall be content with adding some pertinent material, partly general and partly specific. A systematic presentation we leave to future investigators.

The predominance of the spiral tendency is striking in the convolvulus, which from the very beginning is unable to carry on its existence by climbing or creeping, and is compelled to find an upright support for climbing in continual circular movements.

It is precisely this characteristic which provides us with a graphic example and symbol in support of our observations.

Let us slip into a garden during summer to observe a convolvulus, carrying on its growth closely attached to a pole inserted into the ground, climbing in spirals. Now imagine that the pole, like the convolvulus, is alive, that both emerge from the root, react upon one another, and in that way ceaselessly progress. The idea will be easily grasped by anyone who, seeing this image, can turn it into an inner concept: the intertwining plants must acquire from an external source what they ought to, yet fail to, provide for themselves.

The spiral system is more quickly discerned in the dicotyledons. The search for it in the monocotyledons downward will be reserved for later.

We have chosen the twining convolvulus, but many other examples may be found.

Let us now examine that spiral tendency in the little forks, in the vrilles.

---

[95] David Don.

The latter also appear at the ends of composite leaves, where they clearly manifest their tendency to curl.

The completely leafless vrilles themselves must be regarded as branches lacking in solidity; pliable and full of sap, they show a special irritability.

The vrille on the passion flower is self-curling.

Others must be excited and challenged by an external stimulus.

The best example, for me, is the grapevine.

The forklets are extended, searching for contact of some sort; leaning upon any object, they will fasten and clamp themselves to it.

They are branches, like the ones that bear grapes.

Individual grapes are indeed found on the caprioles.

It is remarkable that the third node of the vine produces no vrilles. The explanation is not yet clear to us.

We regard the spiral vessels as the smallest parts that are completely analogous to the whole of which they are a part. Considered as homoeomeriae, they impart their properties to the whole, and receive properties and determinations from the whole in turn. An independent life is given to them, and the power to move individually, on their own initiative, and to assume a given direction. The worthy Dutrochet calls it a vital incurvation. It is not our intention here to investigate these mysteries further.

Let us return to the general subject. The spiral system is terminal and fosters completion.

Specifically, the process occurs by principle and with finesse.

However, it also occurs erratically, prematurely, and destructively.

Our highly esteemed Martius has demonstrated in detail the law whereby the spiral system forms flowers, blossoms, and buds. This law develops directly from metamorphosis, but it required an astute observer to see and to demonstrate it. Let us visualize the flower as a branch moved forward, as a branch curling itself about an axis, buds joined closely and forming a unit. Then we conclude that the buds must make their appearance in circles behind and beside one another, and that they then, simple or diversified, must arrange themselves around each other.

The irregular spiral effect must be regarded as a premature, unfruitful completion. Some stem, branch, or limb is brought to a state where the splint, in which the spiral life is actually operative, increases in predominance, so that the wood or other permanent formation cannot occur.

Let us look at a branch of the ash in the condition described as follows. The splint, which is not held apart by the wood, is crowded together, assuming a flat vegetable appearance. Simultaneously the entire growth contracts, and the buds, instead of developing successively, now are crowded, and eventually appear in unbroken sequence. In the meantime, the whole has been arched, with the remaining wood forming the back, and the result is a most remarkable abnormality of a form turned inward, and resembling a bishop's crook.

Just as the foregoing proves that the life proper of the plant is definitely fostered by the spiral tendency, so too it can be shown that traces of this tendency remain in the completed, enduring part of the plant.

The fresh thread-branches of *Lycium europaeum*, hanging down in full freedom, show only a straight, threadlike growth. When the plant gets older and drier, it clearly tends to wind from node to node.

Even strong trees are seized by such a tendency in their old age; on the Belvedere Chausée,[96] chestnut trees one hundred years old are radically twisted, and the rigidity of the straight-upward tendency is most strangely thwarted.

In the park behind Belvedere Castle, are three slender tall trunks of *Crataegus torminalis*, so clearly twisted from top to bottom, that one cannot fail to note it. We recommend this in particular for observation.

We know instances of flowers, which are folded before unclosing and which develop spirally; there are others which show a twisting during drying.

Pandanus odoratissimus winds spirally from the root up.

Ophrys spiralis winds in such a manner that all the blossoms appear on one side.

The buds of *flora subterranea*, arranged *en échiquier*, lead us to think that they result from a very regular spiral tendency.

In a potato, which was almost a foot long and could hardly be spanned by the hand at its thickest part, it was easy to see the upward spiral sequence of the eyes beginning at the point of attachment and continuing from left to right to the highest point.

In ferns, all growth proceeds from a horizontally placed stem and is directed upward from the sides, both leaf and branch, the latter therefore also supporting and developing the fruit parts. Everything we call fern has its characteristic spiral development. The branches of the horizontal limb are curled into ever smaller orbits, in a double direction, sometimes from the rib spiral and sometimes from the bent-in pinna of the lateral part of the rib, the riblets toward the outside.

See Reichenbach: *Botany for Ladies*, p. 288.[97]

---

[96] In Weimar.
[97] See p. 207.

The birch, without exception, grows spirally upward from the lowermost part of the trunk. Splitting the trunk according to its natural growth reveals the spiral movement from left to right and continuing to the top, a birch 60 to 80 feet in height being turned longitudinally once and even twice. The spiral quality, my cooper tells me, arises when the trunk is more or less exposed to the wind and weather. A trunk standing in the open, for instance near the edge of the forest, exposed especially to the west, will manifest the spiral movement more obviously and pronouncedly than a trunk growing in a thicket. This spiral movement is seen especially in the so-called hoop birch. A young birch, to be used for hoops, is split in the center. If the knife follows the wood, the hoop is useless: for it will turn, as already noticed in the older trunks, once and even twice. For that reason, the cooper needs special tools to cut the birches easily and efficiently. Special tools are needed also for logs of older wood to be used for staves or the like. In cleaving the wood iron wedges must be used that cut rather than split it, for otherwise it would be useless.

That weather, wind, rain, and snow have great influence on the development of the spiral movement, is shown by the fact that hoop birches taken from the thicket are far less subject to spiral movement than those standing alone rather than among bushes and larger trees.

When the spiral tendency came up for discussion last August in Ilmenau, Herr von Fritsch,[98] the huntsmaster, mentioned cases occurring among pines in which the trunk assumed a turning, twisting direction from the bottom upward. Because such trees are found at the forest's edge, it was believed that the external influence of violent storms was the cause. However, the same thing is found even in the densest forests, he said, in approximately one to one and one-half per cent of the trees.

Attention was directed to such trunks for more than one reason. For instance, the wood could not be cut into logs or piled in cords or used for building purposes, since its continual invisible twisting tendency might force the whole structure out of joint.

From the foregoing we see why the curvation continues and increases greatly during the drying of wood, for we have seen quite a number of spiral movements arising and becoming visible solely as a result of the drying process.

The dried pods of *Lathyrus furens* spring apart after complete maturation of the fruit, each curling markedly in an outward direction. When

---

[98] The forester mentioned on pp. 115–116. The discussion alluded to here took place in 1831, a half year before Goethe's death.

such a pod is opened before it is completely ripe, it also shows this spiral direction, though not so strongly and perfectly.

The straight direction of similar plant parts is likewise variously deflected. During a wet summer the pods of sword bean begin to turn, some in snail fashion and others in complete spirals.

The leaves of the Italian poplar have extremely delicate, taut leaf stalks. If pierced by insects, these lose their straightness and assume a spiral direction with two or more twists.

If later such an encased insect swells, the sides of the extended stalks are pressed so close together that they achieve a kind of union, but the case can easily be broken apart at these places and the former shape of the twisted stalk then emerges clearly.

Let us cite the case of the pappus on the seed of *Erodium gruinum*, which holds itself erect, on the support uniting the seeds, until complete maturity and drying, but then forms elastic ringlets, and in the process is tossed about.

Although we have previously declined to treat the spiral vessels as such, we find ourselves compelled to go back even further in elementary botany and refer to *Oscillaria*, the entire existence of which is spiral. Perhaps even more remarkable are the ones bearing the name *Salmacis*, in which the spirals consist of nothing but little contiguous spheres.

However, such suggestions must be alluded to most delicately, if they are to remind us of the eternal harmony of all things.

If the stem of the dandelion is slit at its end and the two sides of the hollow reed gently separated, each will roll in an outward direction and will hang down in a tapering curl. This phenomenon is the children's delight, and by means of it we ourselves can approach closer to one of Nature's deepest mysteries.

These stems are hollow and juicy and can therefore be regarded as splint, and the spiral tendency is inherent in the splint as a living progressing factor. Here, then, we have an example of utmost vertical direction and utmost obscured spiral effort. It might be possible, with microscopic study, to learn more about the interweaving of the vertical and spiral texture.

Vallisneria provides an illuminating example of the way the two systems develop most vigorously side by side, as may be seen from the latest investigations by the custodian of the Royal Botanical Garden at Mantua, Paolo Barbieri. We shall give translations of excerpts from his essays, with remarks of our own interspersed and appended. In this way we hope to carry out our intention more efficiently.

Vallisneria takes root on the bottom of rather shallow, stagnant water, blooming bisexually, in the months of June, July, and August. The male unit appears on an upright shaft, which forms a four-bladed (perhaps three-bladed) sheath at its tip as soon as it reaches the water's surface; in this sheath are found the fruit organs, attached to a conical bulb.

If the stamens are not sufficiently developed, half of the sheath will be empty, and on microscopic observation it will be found that the inner moisture sets about to foster the growth of the sheath and at the same time moves spirally upward in the stalk toward the spadix holding the stamens. In this way the growth and expansion of the spadix is produced simultaneously with the growth of the staminal organs.

By means of this increase in the spadix, the sheath is no longer sufficient to encase the stamens. It therefore divides into four parts; the fruit organs detach themselves by thousands from the spadix, spread and float upon the water, appearing like silver flakes, and make their way to the female plant. The latter rises from the water bed, by releasing the tension of its spiral stem, and on the surface opens a tri-divided crown, in which there are three stigmas. The flakes floating on the water scatter their pollen toward the stigmas and fertilize them. When this has been done, the spiral stem of the female withdraws below the surface, where the seeds, encased in a cylindrical capsule, eventually attain maturity.

All the authors mentioning *Vallisneria* give differing accounts of the manner of fertilization. Some say that the whole complex of the male flower is detached and freed from its short under-water stem by a vigorous movement. Our observer attempted to detach buds from the stem of the male flower and found that none floated about on the water, that they sank to the bottom instead. The structure joining the stem and flower is of greater significance. Here we see no articulation of the kind occurring with detachable plant organs. The same observer examined the silvery white flakes and recognized them as actual anthers; when he observed the spadix to be lacking in all such vessels, his attention was directed to delicate hairs, with a few remaining anthers, on a tri-divided disc, evidently the tri-divided corolla which had enclosed the anthers.

We recommend this strange example, repeated perhaps in other plants, to the serious scientist; but in doing so we cannot refrain from discussing this obvious phenomenon further, though we may repeat ourselves somewhat.

Here the vertical tendency is characteristic of the male. The stem grows straight upward; when it reaches the surface of the water, the sheath develops directly from it, coinciding with it and encasing the spadix, in the manner of calla and others.

Thus we get rid of the current fairy tale about a joint placed most unnaturally between stem and flower, enabling the flower to detach itself and to go hunting lasciviously for a female. The male flower develops only with air and light and their influences, but nevertheless remains firmly attached to its stem. The anthers dart from their stems and float merrily about on the water. Meanwhile the spiral stem of the female relaxes its tension, the flower reaches the water's surface, unfolds, and absorbs the fertilizing influence. The significant change which occurs in all plants after fertilization and which always suggests a degree of rigidity, is also operative here. The spiral tendency of the stem exerts itself, and the stem moves back in the manner it came, following which the seed achieves maturity.

Let us recall the figure of speech we ventured to use with the convolvulus and the staff. Let us go a step further and visualize a vine entwining itself about the elm. Here we see, drawn to our attention by Nature herself, the male and the female, the giver and the receiver together, growing in a vertical and spiral direction.

Let us return now to general aspects and recall what we asserted at the outset, that the vertical and spiral systems are closely bound together side by side in the living plant. When we see that the vertical system is definitely male and the spiral definitely female, we will be able to conceive of all vegetation as androgynous from the root up. In the course of the transformations of growth the two systems are separated, in obvious contrast to one another, and take opposing courses, to be reunited on a higher level.—*Weimar, Autumn of 1831*

On His Plant Studies

Sublime Spirit, Thou gav'st me, gav'st me all
That I sought. Not in vain hast Thou turned
Thy countenance toward me in the flame.
For my kingdom Thou gav'st me Nature,
Power to feel and delight in her wonders;
Thou sendest me, not as a stranger,
But as one who is privileged to search
Deep in the heart of a friend.

—GOETHE, *Faust:* Faust to the Earth Spirit

(English rendering by Aldyth Morris)

The Author Relates the History of His Botanical Studies[1]

Voir venir les choses est le meilleur moyen de les expliquer.—Turpin

To clarify the history of the sciences and to gain accurate knowledge of their progress, we usually make careful inquiry into their origins; we endeavor to determine who it was that first turned his attention to a certain subject; what his procedure was; when and where certain phenomena were first drawn into consideration. In this way new vistas of thought open up from idea to idea, which, universally confirmed through application, finally characterize the epoch wherein what we call a discovery or invention comes into clear view: an approach presenting the most varied opportunity for recognizing and appreciating man's intellectual powers.

The preceding little manuscript[2] has been shown the distinction of having inquiries made into its origin; the wish has been expressed to learn how a middle-aged man of some reputation as a poet, whose time moreover appeared to be taken up by manifold interests and duties, could have ventured into Nature's boundless domain, and studied it so thoroughly that he was able to formulate a theory conveniently stating a rule to be applied to the most varied forms, a rule to which thousands of details must conform.

The author of the brochure mentioned has already given a report on the subject in his morphological notebooks;[3] however, since it is his wish to present all the necessary and appropriate information on this present occasion too,[4] he begs permission to offer here a modest presentation in the first person.

---

[1] The essay as it appears here was published first in 1831. A shorter version called "History of My Botanical Studies" had appeared with "Genesis of the Essay on the Metamorphosis of Plants" (pp. 165–167) in *Natural Science in General; Morphology in Particular*, Vol. I, No. 1 (1817). For Goethe's own translation of the maxim from Turpin, see p. 205.
[2] The essay on the metamorphosis of plants.
[3] See footnote 1, p. 21.
[4] In the second edition of the essay on metamorphosis, 1831.

Born and reared in a large city, I acquired my first schooling in the study of ancient and modern languages, to which rhetorical and poetical exercises were soon added. Supplementing this was everything that would direct an individual to ethical and religious self-examination.

My further education I likewise owe to rather large cities; hence it followed that my intellectual activity was directed toward the manners of polite society and to the pleasant activity which at that time was called "polite literature."

On the other hand, I had no understanding of external Nature in the strict sense of the term, nor the slightest knowledge of her so-called three domains. From childhood on, I had been accustomed to seeing people admire the floral splendor of tulips, ranunculi, and pinks in well-ordered show gardens; and if in addition to the usual fruit varieties, the apricots, peaches, and grapes also turned out well, it was occasion enough for old and young to rejoice. No thought was given to exotic plants, still less to the idea of teaching natural history in the schools.

My first published attempts at poetry were received with approval. However, these first poems described only the inner man and required nothing more than a competent knowledge of human emotions. Here and there might be found a note of passionate delight in rural nature, also a deep urge to apprehend the tremendous secret revealed in constant creation and annihilation; but this urge appears to have been lost in mere uncertain, unsatisfied pondering.

I did not actually enter the life of action or the sphere of science until after my favorable reception[5] into the distinguished Weimarian circle, where in addition to other inestimable advantages I had the joy of exchanging the stuffiness of town and study for the pure atmosphere of country, forest, and garden.

The very first winter afforded the racy, sociable joys of the hunt, resting from which we spent long evenings conversing, not only about all sorts of strange hunting adventures but also, and indeed chiefly, about the necessary cultivation of forests. For the Weimarian forestry commission had excellent men, among whom the name Sckell[6] is of blessed memory. A revision of all the forest preserves, based on surveying, had already been completed—plans for annual tree-fellings had for years been made long in advance.

This sensible course was followed with interest even by the younger members of the nobility, of whom I shall here mention only Baron von

---

[5] In 1775.
[6] See p. 188.

Wedel,[7] unfortunately torn from us in his best years. He handled his affairs in a straightforward way and with great good sense. Indeed, even at that early date he was urging the reduction of the game supply, convinced that its conservation would be detrimental not only to agriculture but also to forestry.

Here in Weimar the forest was revealed to us in its full length and breadth. Not only the lovely ducal estates themselves but also the adjoining estates were, by virtue of pleasant neighborly relations, accessible to us, and all the more so as the rising science of geology, with the ambition of youth, was attempting to give an account of the ground and soil upon which these ancient forests had taken root. Conifer forests of all kinds, with their somber greenness and balsam fragrance, beech groves of more joyful appearance, the slender birch and the low, nameless underbrush, each had sought and won its place. We could survey all this in more or less well-forested regions extending for miles.

Then when the practical utilization of tree varieties arose for discussion, inquiries into their qualities had also to be made. The practice of tapping trees for resin, the abuse of which authorities gradually sought to restrict, led to an examination of the fine balsam juices associated with such trees from root to crown during two centuries of growth, nourishing them and keeping them eternally green, fresh, and alive.

Here, too, we saw the entire family of mosses in greatest diversity. We even studied the roots hidden below the earth. For to those forest regions, from remotest times onward, there had been a migration of herbalists who, working with secret formulas passed from father to son, had prepared many kinds of extracts and spirits; and the universal reputation of these latter for possessing curative powers was renewed, extended, and utilized by assiduous workers called balsam-carriers. In these activities the gentian played a big role, and it was a pleasant task to examine closer the plant and flower of this rich genus in its various forms, especially the salutary root. This was the first plant genus to attract me in earnest and later to inspire me to continue my study of its species.

Here it may be noted that my botanical education resembled to a certain degree the course of botanical history itself, for I had progressed from superficial observation to useful application, from need to knowledge. And what botanical expert perusing the above will not smilingly remember the era of the rhizotomes?[8]

---

[7] Otto Joachim Moritz von Wedel, chamberlain and chief forester of Weimar.
[8] The rootdiggers of old Greece, with whom the recorded history of botany began.

However, since at present it is my intention to record only my approach to the scientific aspect of botany, I feel impelled first of all to do homage to the memory of a man deserving in every respect of the high esteem of his fellow Weimarians. Dr. Buchholz,[9] owner of the only pharmacy in Weimar at that time, well-to-do and filled with zest for living, with praiseworthy intellectual curiosity directed his activity to the natural sciences. For pharmaceutical work he sought out the most efficient chemical assistants. Indeed, the excellent Göttling[10] left this laboratory as an accomplished chemist. Every noteworthy new discovery in Germany or abroad was tested here under the direction of the laboratory chief and altruistically presented to a circle of intellectually curious friends.

Later also—if in his honor, I may be permitted to get ahead of my story—when the scientific world was zealously engaged in studying various gases, he never failed to show us, through experiments, the newest developments. For instance, he had one of the first Montgolfian balloons released from our terraces—to the delight of the initiated, although the mass of people were so astonished they could scarcely contain themselves, and hosts of frightened doves took refuge in flight.

At this point perhaps I shall have to anticipate the reproof that I am bringing extraneous matter into my essay. Permit me to reply to this that I could not write coherently of my education if I did not recall with gratitude my early advantages in the Weimarian circle, so intellectually advanced for its time, where taste and knowledge, science and poetry endeavored to work together in harmony, where serious, solid study vied constantly with happy, lively activity.

More closely considered, moreover, what I have just said does have a connection with my subject. For at that time, the sciences of chemistry and botany were jointly developed as the result of the needs of medical science, and just as the famous Dr. Buchholz ventured out from his pharmacopia into higher chemistry, so too did he advance from his restricted herb gardens into the larger plant world. In his gardens he undertook to raise not only such plants prescribed for pharmaceutical purposes but, in addition, rare and newly discovered plants for scientific purposes.

The young regent,[11] who had early devoted himself to the sciences,

---

[9] Wilhelm Heinrich Sebastian Buchholz, 1734–1798.
[10] Johann Friedrich August Göttling, 1755–1809, professor of natural sciences in Jena.
[11] For the part played by Duke Karl August of Weimar in the cultivation of plants in Weimar, see pp. 186–190.

directed this man's activity toward more general practice and study, converting large sunny garden areas in the vicinity of shady and moist places into a botanical establishment. In this project older and experienced court gardeners immediately offered their assistance. Those catalogues of this establishment still extant testify to the ardor with which such projects were carried on.

Under the circumstances I, too, was obliged more and more to seek illumination in matters botanical. Linné's *Terminology*, his *Fundamentals*[12] upon which the great structure was to rest, Johann Gessner's *Dissertations in Explanation of Linnean Elements*,[13] all bound in a single slender volume, accompanied me into the highways and byways, and today that same volume reminds me of the active, happy days when those precious pages opened up a new world to me. Linné's *Philosophy of Botany*[14] I studied daily, thus advancing farther and farther in ordered knowledge, attempting to acquire as far as possible all that might procure for me a more general view of this broad realm.

Especially advantageous, as in all matters scientific, was the proximity of the Jena Academy, where medicinal plants had long been seriously and industriously cultivated. In addition, Professors Praetorius,[15] Schlegel,[16] and Rolfink[17] had already acquired the esteem of their contemporaries in the realm of general botany. Epoch-making, moreover, was Ruppe's *Flora Jenensis*,[18] which had appeared in 1718; for thereafter the plant studies, previously limited to a narrow cloister garden, were extended to the entire rich region, introducing a free, joyous study of Nature.

Intelligent country folk in the vicinity, who had previously assisted chemists and herb sellers[19] and had little by little succeeded in learning the newly introduced terminology, participated zealously in the project. In Ziegenhain the Dietrich family distinguished itself especially. The head of the family[20] had even attracted the attention of Linné and could display a holographic letter from this highly esteemed man—a diploma,

---

[12] Linné's *Termini botanici* and *Fundamenta botanices*.
[13] Johann Gessner, 1709–1790, professor in Zürich.
[14] Linné's *Philosophia botanica*. The original edition, Stockholm, 1751, was in Goethe's possession.
[15] Hieronymus Praetorius, 1595–1651, at one time professor of physics at Jena.
[16] Paul Marquard Schlegel, 1605–1653, professor of medicine and director of the botanical garden in Jena.
[17] Werner Rolfink, 1599–1673, after 1629 professor of anatomy and botany and director of the old botanical garden in Jena.
[18] Heinrich Bernhard Ruppe, 1689–1719.
[19] The herbalists mentioned on p. 151.
[20] Adam Dietrich, 1711–1782.

as it were, through which he justifiably felt elevated to the botanical nobility. After his death, his son took over the business, which consisted chiefly in furnishing teachers and students with so-called lessons, that is, bundles of plants blossoming each week in the vicinity. The activities of this jovial man extended as far as Weimar, thus enabling me to become acquainted with the abundant flora of Jena.

A still greater influence on my course of study was exerted by a grandson, Friedrich Gottlieb Dietrich.[21] A well-built youth of regular, pleasant features, he strode forward[22] with fresh, youthful vigor to master the plant world. His brilliant memory which held fast all the unusual terms could instantaneously produce them. I liked to have him about me, for his nature and behavior revealed a frank, independent character, so I decided to take him with me on a trip to Karlsbad.

In hilly regions,[23] always on foot, he zealously tracked down growing things, whenever possible handing them into my carriage on the spot, and, in the manner of a herald, announcing the Linnaean designations, both genus and species, with happy conviction, if sometimes with the wrong pronunciation.[24] In this way I attained a new relationship to open, splendid Nature, while my eyes enjoyed her wonders, and at the same time the scientific designations of the individual plants reached my ears as though from a distant study chamber.

In Karlsbad the energetic young man was in the hills at sun-up, bringing abundant "lessons" to me before I had emptied my first mug at the spa. The hotel guests[25] all participated, especially those who themselves pursued this beautiful science. They found their minds stimulated in the most charming way by the sight of a handsome jerkin-clad country boy, running about, exhibiting great bundles of plants and designating them by names of Greek, Latin, and barbaric origin. It was a phenomenon that excited much interest among the men—and apparently also among the women!

Lest the aforesaid appear much too dilettantish to the professional man of science, let me say at once that it was exactly this lively behavior that won for us the favor and interest of a man who was rather well-

---

[21] 1768–1850.
[22] On an excursion with Goethe to the Hausberg, a mountain near Jena, in the summer of 1785.
[23] The journey led past the Fichtelgebirge (Fir Mountains) in middle Germany.
[24] For Goethe's dislike of incorrect pronunciation of Latin plant names, see p. 190.
[25] Among those staying at Karlsbad at that time, were the Duchess Luise, Frau von Stein, Herder, Count and Countess Brühl, Princess Lubomirska, Countess Bernstorff.

versed in this field, an excellent physician who, accompanying a prominent and distinguished man, had planned to spend his time at the spa in botanical pursuits. He soon joined us and we were happy to lend him a helping hand. He endeavored to preserve most of the plants that Dietrich brought in early in the morning, in the process writing down names and many other things besides. From this practice I could derive only benefit. Through repetition the names were engraved in my memory, and I gained greater skill in analysis—without conspicuous success however, for I was by nature averse to classification and counting.

Of course, our busy endeavors and activities also had several opponents among the distinguished visitors. We repeatedly heard it said that this science of botany which we were so assiduously pursuing was by and large only a nomenclature, a system based on counting—and not very accurate counting at that; that it could satisfy neither the reason nor the imagination, and that it could achieve no satisfactory results. In spite of this objection we confidently pursued our way, which indeed promised to take us far enough into the science of plants.

At this point I should merely like to add briefly that young Dietrich's subsequent career was in harmony with these beginnings. He strode forward untiringly on this path, gained fame as a writer,[26] and had a doctorate conferred upon him. To this day he zealously and creditably heads the grand ducal gardens in Eisenach.

August Wilhelm Batsch,[27] son of a father universally beloved and esteemed in Weimar, had spent his student years in Jena profitably, devoting himself to the natural sciences and attaining such good results that he was called to Köstritz to classify Count Reuss's large collection of natural history specimens and to be its director for a while. Then he returned to Weimar, where, during the severe winter, so hostile to plant life, I had the pleasure of making his acquaintance at the skating grounds —at that time a gathering place of good society. I very soon learned to value his gentle determination and quiet zeal; and while indulging in active, out-of-door sport I discoursed openly and at length with him regarding advanced views of botany and the various methods of applying this knowledge.

His manner of thought was highly suited to my own wishes and needs, for what he was aiming at was the arrangement of plant families in ascending and gradually developing progression. This natural method,

---

[26] Among other works he wrote the large *Lexicon of Gardening and Botany*.
[27] See pp. 76, 96, 183, 189, 196, 217.

which Linné himself, entertaining pious hopes, had pointed out,[28] and at which French botanists persevered theoretically and practically, was now to occupy for life a younger, more enterprising man. How happy I was to have my share at first hand!

However, I was to be immeasurably assisted not alone by these two youths, but also by a worthy man of advanced years. Aulic Councillor Büttner[29] had brought his library from Göttingen to Jena and I was commissioned by my duke, who had procured this treasure for himself and for us, to initiate preparations for setting it up in accordance with the plan of the owner, who was retaining possession of his collection. Thus I was in constant intercourse with this man who was a walking library, prepared to give detailed and satisfying answers to every question and preferring above everything else to discuss botany.

In these discussions he did not deny, indeed, he passionately avowed that he had never accepted the system of his contemporary, Linné, the distinguished man whose fame had spread throughout the world; that in quiet opposition he had endeavored to arrange the plants according to families, advancing from the simplest, almost invisible rudimentary manifestations to the most complex and devious. He was fond of exhibiting an outline, delicately written in his own hand, in which the species were arranged in this manner, greatly to my edification and satisfaction.

No one who meditates upon the foregoing will fail to recognize the advantages that my position afforded for investigations of that kind: great gardens in the city as well as in country estates; various plantings, in the vicinity, of trees and bushes that had been undertaken not without botanical consideration; availability of local flora that had long been scientifically investigated; and moreover the influence of a progressive academy—all these things together offered assistance enough to gain insight into the plant world.

While my knowledge and views of botany were being expanded in this sociable companionship, I also came upon a hermit-like friend of plants, one who had earnestly and industriously devoted himself to this field. And who would not wish to follow the revered Johann Jacob Rousseau[30] on his lonely wanderings, where, at odds with mankind, he turned his attention to the plant and flower world and with true, forth-

---

[28] The so-called natural system of plants, which Linné himself had designated as the most important future work of botany. Pioneer work in it was carried on by Bernard and Antoine de Jussieu, definitively by the latter in 1789.

[29] Christian Wilhelm Büttner, 1716–1801, professor in Göttingen, later a private scholar in Jena.

[30] Jean Jacques Rousseau.

right vigor of intellect made himself acquainted with the mute and attractive children of Nature!

I do not know whether in his early life he had any interest in flowers and plants aside from that deriving from sentiment, taste, and tender memories. According to his express remarks, however, he may not have become aware of this realm of Nature in all its fullness until after a stormy literary life on St. Peter's Island in Lake Bienne.[31]

Later on in England,[32] one observes, he already had a more independent and extensive view of it. His relationship to plant lovers and connoisseurs, especially to the Duchess of Portland,[33] may have directed his discerning eye to a more and more extensive area; and an intellect such as his, which felt impelled to prescribe law and order to nations, must inevitably arrive at the conclusion that such great diversity of forms in the boundless world of plants would be unthinkable without a fundamental principle, be it ever so concealed, to restore uniformity. He engrosses himself in this realm, absorbs it earnestly, feels that accurate methodical progress through the whole is possible, does not trust himself, however, to step forward with it. It is always a pleasure to listen to his own explanation of the matter:

"As far as I am concerned, I am only a student in this field and am not thoroughly grounded in it. When I botanize, I am thinking more of diversion and pleasure than of instruction, and I cannot in my hesitant observations presume to instruct others in a field I myself do not know.

"Yet I confess that the difficulties I encountered in my study of plants caused me to arrive at several methods whereby the study might be made easier and beneficial to others, by following the thread of a plant system by a method more progressive and less removed from the senses than the one pursued by Tournefort[34] and all his followers, Linné not excepted. Perhaps my plan is not feasible; we shall discuss it when I have the honor of seeing you again."

Thus he wrote at the beginning of the year 1770.[35] However, in the meantime the subject pursued him; already in August, 1771, we

---

[31] In 1765.

[32] In 1766.

[33] Rousseau corresponded with the Duchess of Portland on the subject of botany during 1766–1767.

[34] Joseph Pitton de Tournefort, 1656–1708, professor of botany in Paris, famous traveller, inventor of a system of plants before Linné. His system was based on the structure of the corolla.

[35] In a letter to his friend in Lyons, De la Tourette.

find him undertaking at the request of a friend[36] the duty of teaching others, indeed of lecturing to women on what he knows and understands, not for the purpose of entertaining them but of giving them a solid introduction to science.

Here he succeeds in tracing his knowledge back to the first physically demonstrable elements. He presents the plant parts, and teaches how to distinguish and name them. First he assembles the complete flower again from the individual parts, designating them either by popular names or by genuine Linnaean terminology. Then he immediately gives a broader view of whole masses of them. Gradually he introduces us to lilies, legumes and podlets, labiate and mask flowers, umbels, and composites; at the same time that he is thus clarifying the differences in ascending diversity and interlacement, he guides us imperceptibly to a complete and gratifying perspective. For since he is dealing with females he manages, discreetly and suitably, to point out utility, advantages and disadvantages; and he does this all the more skillfully and easily because he takes examples from the immediate vicinity, speaking only of indigenous plants and not presuming to consider exotic ones, even those known and cultivated domestically.

Under the title, *Rousseau's Botany,* all his writings on the subject were nicely published in 1822 in small folio, with colored illustrations, in the manner of the worthy Redouté,[37] of all the plants he had discussed. In examining these illustrations one notes with pleasure that in his studies he took what was at hand in his native environment, presenting only such plants as he himself could observe on his walks.

His method of narrowing down the plant world lends itself to the classification of plants according to families, as we have seen above; and since I too at that time had been led to conclusions of this kind, I was all the more forcibly impressed by his presentation.

And just as young students prefer young teachers, so also does the dilettante like to learn from the dilettante. This would of course be questionable from the standpoint of thoroughness if experience did not show that dilettantes contribute a great deal to science. This is quite natural: the professional man, striving for completeness, investigates a wide field extensively; with the amateur it is a matter of succeeding

---

[36] Madame Delessert, a married daughter of De la Tourette, turned to Rousseau for advice in the botanical instruction of her little daughter, whereupon Rousseau wrote her eight letters during 1771–1773 on the method of learning and teaching botany. Goethe read these letters in the summer of 1782 in Rousseau's works, which had appeared in the meantime.

[37] Pierre Joseph Redouté, 1759–1840, flower painter.

with a single portion of the field and of attaining a height whence he may have a view, if not of the whole, at least of its greater part.

I shall cite only so much of Rousseau's work as to say that he exhibits a charming solicitude for the drying of plants and the arrangement of herbaria, sincerely mourning the loss of his own; but even in this, at odds with himself, he probably had neither the skill nor the perseverence to give much attention to the preservation of plants on his many excursions and for that reason no doubt likes to refer to such collections as "hay."[38]

When, however, as a favor to a friend, he treats the mosses with proper attention, we recognize most forcefully the profound interest which the plant world has elicited from him, an observation completely confirmed by the *Fragments pour un dictionnaire des termes d'usage en botanique*.[39]

Let us here mention only enough to indicate how much we owed to him in that epoch of our studies.

As we thus behold him now, freed from all nationalistic obstinacy, adhering to Linnaean methods—irresistible in their progress—we may remark, for our part, that is a great advantage, when entering what is to us a new scientific field, to find it in a state of crisis and to find an extraordinary man turning that crisis to advantage. We are then young with the young method; our initiation into the science coincides with a new epoch; and we are drawn into the group of pioneers as into a natural element that buoys us up and carries us along.

Thus, with the rest of my contemporaries, I had become aware of Linné, of his farsightedness and of his compelling authority. I had devoted myself to him and to his theory with complete trust. Nevertheless, I gradually became aware that some things on the path which he had marked out and I had taken, were holding me back, if not actually leading me astray.

If I am to become consciously articulate about these circumstances, let the reader think of me as a born poet, who, in order to do justice to his subjects, always seeks to derive his terminology directly from the subjects themselves, each time anew. Imagine that such a man is now expected to commit to memory a ready-made terminology, a certain number of words, and bywords, with which to classify any given form, and by a happy choice to give it a characteristic name. A procedure of that sort always seemed to me to result in a kind of mosaic, in which

---

[38] Rousseau sold his first big herbarium in 1775 because it was a burden to him on his journeys and in the small apartments he lived in.

[39] Published in 1782, after Rousseau's death.

one completed block is placed next to another, creating finally a single picture from thousands of pieces; this was somewhat distasteful to me.

To be sure, I recognized the necessity of this procedure, which had as its goal the discussion of certain external plant phenomena, according to general agreement, and the elimination of all phenomena that are uncertain and difficult to represent. Nevertheless, when I attempted an accurate application of terminology, I found the variability of organs the chief difficulty. I lost the courage to drive in a stake, or to draw a boundary line, when on the selfsame plant I discovered first round, then notched, and finally almost pinnate stems, which later contracted, were simplified, turned into scales, and at last disappeared entirely.

The problem of designating the genera with certainty, and of arranging the species under them, seemed insoluble to me. Of course, I read the method prescribed, but how could I hope to find a suitable classification when even in Linné's time genera had been shattered and separated, and classes themselves dissolved? The conclusion to be derived from all this seemed to be that even this highly astute man of genius had been able to subjugate Nature only in a general way. My admiration for him was not in the least reduced through this; nevertheless, a very special conflict was bound to arise. The reader can imagine my embarrassing situation, a self-taught tyro torturing himself and fighting his way through.

However, it was necessary that I pursue without interruption my other duties and recreations, which by good fortune took me out of doors. Thus it was through direct observation that my attention was powerfully drawn to the following fact: each plant seeks its own advantage and demands conditions in which it may grow in fullness and freedom. Mountain height, deep valley, light, shade, aridity, moisture, heat, warmth, cold, frost, whatever other conditions there may be! Plants require genera and species in order to grow with full strength and abundance. To be sure, in certain places and in many situations, they yield to Nature and allow themselves to be swept into variations, without however giving up completely the form and quality acquired through their own efforts. Such thoughts came to me in the great out-of-doors, shedding new light for me on gardens and books.

If a connoisseur should take a notion to transport himself back to the year 1786, he might possibly be able to picture the situation in which I had been caught for ten years, although it would be quite a problem even for a psychologist, inasmuch as my obligations, interests, duties, and diversions would all have to be included in that representation.

Here the reader must allow me to insert a remark that has a bearing upon the whole: everything that has been round about us from youth, with which we are nevertheless only superficially acquainted, always seems ordinary and trivial to us, so familiar, so commonplace that we hardly give it a second thought. On the other hand, we find that new subjects, in their striking diversity, stimulate our intellects and make us realize that we are capable of pure enthusiasm; they point to something higher, something which we might be privileged to attain. This is the real advantage of travel and each individual benefits in proportion to his nature and way of doing things. The well-known becomes new, and, linked with new phenomena, it stimulates attention, reflection, and judgment.

In this manner my interest in Nature, especially the plant world, was vigorously stimulated by a quick journey across the Alps. The larch tree, more common there than elsewhere, and the cembra pine tree, a new phenomenon, drew my attention strongly to climatic influence. Other plants, more or less modified, did not remain unnoticed during my hasty trip. However, the luxuriance of the foreign vegetation impressed me most forcibly when I visited the Padua Botanical Garden,[40] where a high, broad wall with the fiery red bells of the *Bignonia radicans*[41] glowed enchantingly before me. In addition, I saw in the open many rare plants that previously I had seen growing only in hothouses. By this time also some of the plants standing in the open, enjoying the benefits of fresh air, were lightly covered to protect them against the more severe season. A fan palm attracted my attention; fortunately, the simple, lance-shaped first leaves were still near the ground; the successive separation increased until finally the fan quality was discernible in complete development. From a spatulate sheath, a branchlet with blossoms finally emerged, looking like an old offspring, strange and surprising, and unrelated to the preceding growth.

At my request the gardener cut off an entire sequence of modifications for me, and I burdened myself with several pasteboard containers in which to carry these treasures around. I still have them, in good condition and just as they were when I first took them away with me, and I value them as symbols capable of attracting and engaging my attention completely, a flourishing reward for my efforts, so to speak.

The variability of plant forms, whose unique course I had long been following, now awakened in me more and more the idea that the plant forms round about us are not predetermined and established; instead, we find allotted to them, along with their stubborn clinging to genera and species, a happy mobility and flexibility, enabling them to adapt

---

[40] September 26, 1786.
[41] The plant, however, was the Japanese *Bignonia grandiflora*.

themselves to the many conditions throughout the world which influence them, and to be formed and reformed in accordance with them.

Here variations in soil come into consideration; richly nourished by valley moisture, stunted by the aridity of heights, entirely protected against frost and heat or inescapably exposed to both of them, the genus can be modified to the species, the species to the variety, and the latter in turn to other varieties ad infinitum; and at the same time the plant is restricted to its own realm, even when it attaches itself in neighborly fashion to the hard stone, or to more animated life here and there. But even the most distantly related ones have a marked affinity and permit easy comparison.

Because they may be grouped under one concept, it gradually became clear to me that the concept could also be valid in a higher sense: a challenge which hovered in my mind at that time in the sensuous form of a supersensuous plant archetype. I traced the variations of all forms as I came upon them. In Sicily, the final goal of my journey, the conception of the original identity of all plant parts had become completely clear to me; and everywhere I attempted to pursue this identity and to catch sight of it again. In this way there was engendered in me an interest, a passion, which affected all the necessary and optional transactions and pursuits of my return journey. Only a person who has himself experienced the impact of a fertile idea, be it original, or communicated to him, or engrafted by others, will understand what passionate activity is stirred in our minds, what enthusiasm we feel, when we glimpse in advance and in its totality something which is later to emerge in greater and greater detail in the manner suggested by its early development. Thus the reader must surely agree that, having been captured and driven by such an idea, I was bound to be occupied with it, if not exclusively, nevertheless during the rest of my life.

Even though this interest had taken a strong hold on my mind, regulated study on my return to Rome was not to be thought of; poetry, art, and antiquity all seemed to demand my entire attention, and seldom in my life have I experienced more arduous and laborious days. Professional men will perhaps consider me too utterly naive when I tell how day after day, on walks and little excursions, I laid hold of all the plants I caught sight of in the gardens. During seed time it seemed especially important to me to observe the manner in which some of the seeds emerged into the light of day again after having been placed in the soil. Thus I observed the sprouting of *Cactus opuntia*,[42] which is irregular in

---

[42] The so-called Indian fig, the wild American cactus, which is to be observed everywhere in southern Italy.

its growth, and saw with pleasure that it made its appearance as an innocent dicotyledon with two tender leaves, and only developed the irregularity during its future growth.

I also had a remarkable experience with seed capsules. I had taken home several of *Acanthus mollis* and had put them into an open box. One night I heard a crackling and soon after a noise as though little objects were jumping about on the walls and ceilings. I could not explain it immediately, but later I found that my pods had burst and had strewn the seeds round about. The dryness of the room had matured them to this degree of elasticity in a few days.[43]

I must mention several others of the many seeds that I observed in this way, for, as a memorial to me, they continued to grow for some time in old Rome. Pine kernels expanded very remarkably: they arose as if enclosed in an eggshell but soon threw off this cap and exhibited the beginnings of their future destiny in a wreath of green needles.[44] Before my departure I had already planted the rudiment of a future tree, already of some size, in the garden of Madame Angelica,[45] where it flourished many years and reached a considerable height. Interested travelers used to tell me about it, to our mutual delight. Unfortunately, the person who took over the property on her death, thought it odd to have a pine tree growing in the wrong place among his flower beds and forthwith banished it.

Several date plants which I had raised from seeds were more fortunate, and indeed I was able to observe the development of several specimens. I turned them over to a Roman friend, who planted them in a garden, where, as a traveling nobleman[46] graciously assured me, they are still flourishing. They have grown to a man's height. Let us hope they will not become burdensome to the owner, and will continue to thrive and flourish!

The preceding concerns propagation by seeds; however, through Privy Councillor Reiffenstein[47] I became just as interested in propagation by buds. On our walks he would tear off branches here and there, insisting to the point of pedantry that if they were stuck into the ground they would grow immediately. For decisive proof he would point to slips of that kind in his own garden. How important and widespread such methods of propagation have become in commercial-botanical gardening, a circumstance which I wish he had lived to see!

[43] Cf. par. 78, pp. 62–64.
[44] Cf. par. 16, p. 37.
[45] The painter, Angelica Kauffmann, Goethe's friend in Rome.
[46] Ludwig I, of Bavaria.
[47] Johann Friedrich Reiffenstein, 1719–1793.

Most remarkable to me, however, was a carnation plant that had grown to such an extent as to resemble a bush.[48] The great vitality and propagating power of this plant is well known; bud upon bud is crowded on its branches and node funnelled into node. In this case, the process had been intensified through longevity and the buds had been forced from an unascertainable circumscription to the highest possible development, so that even the completed flower again produced four perfect flowers from its heart.

As I had no means of preserving this miracle-plant, I undertook to sketch it in detail, in the process gaining more and more insight into the basic conception of metamorphosis. However, the division of interest caused by my many obligations became more and more of a hindrance, and my sojourn in Rome, whose end I could now foresee, became more and more painful and burdensome to me.

On my return journey I pursued these thoughts incessantly. In my mind I planned a suitable exposition of my views, wrote it down soon after my return, and had it printed.[49] It appeared in 1790, and I had the intention of following it soon with an additional exposition and the necessary illustrations. My busy life, however, interrupted and thwarted my good intentions, and for these reasons it pleases me all the more that the present reprinting of my essay now provides occasion for recalling my participation of more than forty years in these pleasant studies.

Up to this point I have tried to outline as clearly as possible the way I pursued my botanical studies, into which I was guided, driven, forced—and then held captive by my interest—studies to which I devoted a considerable part of my life. However, some otherwise friendly reader might possibly rebuke me, saying I have tarried too long with details and individual personalities; for that reason I should like to explain here that I have done so purposely and not without design, with the intention of adding some general remarks after giving the details.

For more than half a century I have been known as a poet, in my own country and undoubtedly also abroad; or at any rate I have been permitted to pass for one. But the fact that I have busily and quietly occupied myself with Nature in all her general physical and organic phenomena, constantly and passionately pursuing seriously formulated studies—this is not so generally known; still less has it been accorded any attention.

When my essay, printed forty years ago in German, with its ingenious explanation of the laws of plant formation, became better known in

---

[48] See par. 105 and par. 106, pp. 73–74.
[49] The first edition of the essay on metamorphosis.

Switzerland and France, people were extremely astonished to find that a poet, who normally occupied himself with moral phenomena and specifically those associated with feeling and power of imagination, could turn for a moment from his path and in a cursory study achieve such an important discovery.[50]

It is to combat this mistaken idea that the present essay has been written; it is intended to make clear how I found opportunity to devote a great part of my life, with interest and passion, to nature studies.

It was not through extraordinary intellectual gifts, not through momentary inspiration, not unexpectedly or suddenly, but through logical effort that I arrived at such a satisfying result.

To be sure, I might have complacently accepted the high honor people intended to pay to my sagacity, or at least have taken some secret pride in it. However, as it is equally harmful in scientific pursuits to rely on either experience or reason exclusively, I consider it my duty to record the event for serious investigators just as it occurred, historically accurate, though not in complete detail.

Genesis of the Essay on the Metamorphosis of Plants

THROUGHOUT THE FOREGOING[51] I had not ceased to go forward along the path marked out by Linné, upon which, however, I found a good many things holding me back if not actually leading me astray. I conscientiously attempted to apply botanical terminology to plant parts, but unfortunately was very greatly impeded in the process. For instance, when on the selfsame stem I saw what was indubitably a leaf gradually turning into a stipule;[52] when on the selfsame plant I discovered first rounded, then notched, and finally almost pinnate leaves, I lost the courage to drive in a stake or even to draw a mere line of demarcation. The hardest task for me was to classify the genera and to subordinate the

---

[50] This sentence is directed against a well-intentioned statement of the great botanist Augustin Pyrame de Candolle, 1778–1841, to the effect that Goethe's discovery of metamorphosis resulted from a mere side glance of the poet into a strange field. See pp. 10, 205, 206 for direct references to de Candolle.

[51] The events related in the original, brief sketch for "The Author Relates the History of His Botanical Studies." The latter had first appeared in *Natural Science in General; Morphology in Particular,* Vol. I, No. 1 (1817), together with the present essay, and was then extended for the 1831 edition of the *Metamorphosis.* See footnote 1, p. 149.

[52] By the term Goethe here understands a much reduced leaf formation contracted to form a scale.

species with certainty. The method prescribed I was familiar with, of course, but how could I hope to make a reliable application when, since Linné's time, so many a plant family had been further cleft and rent asunder? From this circumstance it seemed to me that even the most experienced and sharp-sighted man would find it hard to come to terms with Nature. Likewise the conflict in which the varieties and sports were engaged with the species themselves caused me to ponder. Who would deny that under certain circumstances abundant nourishment changed the supposed character of a plant completely? And what was one to make of the many irregular malformations?

I therefore felt justified in concluding that Linné and his successors had proceeded like legislators, less concerned with what is than with what should be, giving no consideration to the nature and requirements of individual citizens, but intent rather upon solving the difficult problem of how so many unruly, inherently unfettered beings can be made to exist side by side with a degree of harmony.

Studying Linné's undertaking from this point of view, as presented in the much lauded and much loved booklet,[53] I felt more and more reverence for this extraordinary man, more and more respect for his successors, who did not allow the reins seized by him to drop from their hands but managed to retain control in Linné's characteristic style.

A brief, calm glance was sufficient to make me realize it would require a lifetime to gain a panoramic view and to bring order into the infinitely free vital activity of one single natural realm, even granting that innate talent justified and inspired the undertaking. At the same time I felt that for myself there might exist another way, analogous to my own way of life in general. The phenomena of formation and transformation had taken a mighty hold on me; here Nature and the power of imagination seemed to be vying with one another to see which could proceed with greater boldness and to greater lengths.

I became more and more aware of such flexibility; during my travels, particularly, I traced it in different geographical latitudes, under varying barometric pressures, etc. Of all this, the notes I had begun to produce during my Italian journey provide a preliminary account. The next volume will describe how, quite naively, I first conceived the idea of plant metamorphosis, pursuing it with joy, ecstasy indeed, lovingly immersing myself in it in Naples and Sicily, applying it everywhere, and writing reports to Herder on what had taken place, with as much enthusiasm as was manifested over the finding of the lost silver piece in the

---

[53] See p. 153.

gospel parable. In the planned book there will be a detailed report of all this.

In the same way I shall trace out in greater detail the observations I made during my second sojourn in Rome, of the lush and luxuriant vegetation propagating itself in so many different ways. I shall describe many an hour spent studying and sketching a number of proliferations, of rare occurrence in our own climate; I shall describe how the theory was traced out in my mind during this process, much in the form it was later expressed. From my diaries I shall give a sufficient account of all this, unconcerned whether anyone should impudently venture to use these pure confessions as arguments against me, as evidence of my errors, as has unfortunately happened in other cases, to spoil with cant and verbiage, for me and for others, the independent, happy pursuit of Nature's truth.

Finally impelled to return to Germany, thrust irrevocably from the world of art, given over to despair, I felt all the more vividly the value of the world of Nature. In search of healing and comfort, I passionately resumed all former pursuits that might reunite me with natural science and its friends. One of the first pieces of work was the essay here reprinted.[54] Well known, and after almost thirty years finally accepted in scientific circles, it is herewith recommended anew to the grace and favor of friends and students of Nature's unceasing activity.

The History of the Manuscript[55]

From Italy, rich in forms, I was plunged back into formless Germany, exchanging a sunny sky for a gloomy one. My friends, instead of comforting me and drawing me back to them, drove me to despair. My delight in things remote and almost unknown to them, my sorrow and grief over what I had lost, seemed to offend them. I received no sympathy, no one understood my language. I could not adjust myself to this distressing situation, so great was the loss to which my exterior senses must become reconciled. But gradually my spirit returned and sought to preserve itself intact.

Without interruption, for two successive years,[56] I had been constantly observing, collecting, and pondering, in an attempt to round out each

---

[54] The essay on the metamorphosis of plants.
[55] Written in April, 1817; published in *Natural Science in General; Morphology in Particular,* Vol. I, No. 1 (1817).
[56] From autumn, 1786, to summer, 1788.

of my undertakings. Having gained some insight into the way the gifted Greeks had developed a supreme art within their own national sphere, I had hope of gradually surveying the whole field of art and of acquiring for myself a pure and unprejudiced enjoyment of it. In addition, I prided myself on understanding Nature's method in producing, in accord with definite laws, a living structure that is a model of everything artistic. The third subject on which I was working involved national customs. My purpose in this study was to discover through the clash of inevitability and accident, compulsion and desire, impetus and inertia, the manner in which a third thing emerges that is neither art nor nature, but both at the same time—inevitable and accidental, premeditated and fortuitous. Human society is something I understand well.

As I moved about in these fields, intent on developing my knowledge, I undertook to put down in writing the concepts that were most clearly formulated in my mind; in this way my thinking was regulated, my experience organized, and fugitive thoughts held fast. I was simultaneously writing an essay on art, fashion, and style, one on the metamorphosis of plants; and "The Roman Carnival."[57] Each of these reveals what was going on in my mind at the time and the position I had taken with respect to those three great fields of knowledge. I finished first the *Essay on the Metamorphosis of Plants*, which traces the manifold specific phenomena in the magnificent garden of the universe back to one simple general principle.

Now it is a well established fact that we writers are pleased with what we write, for we should otherwise not have taken the trouble to write it in the first place. Well satisfied with my brochure, I was flattered to believe myself auspiciously launched on a career in science. However, the same thing happened to me that I had experienced in my purely literary work; again, at the very outset I was repulsed; however, in this latter case the first obstacles likewise foretold the later ones; and even down to the present day I live in a world from which I can communicate scientific information to only a few people. But to get back to the point, my manuscript had the following history.

I had every reason to be satisfied with Herr Göschen, the publisher of my collected works, but unfortunately their publication came at a time when Germany had forgotten me and did not care to be reminded of me; at least I gained the impression that my publisher did not find their sale up to his expectations. However, I had promised to offer future manuscripts to him before offering them to anyone else, an arrange-

---

[57] A chapter from Goethe's Italian letters and journals, published separately, illustrated with etchings of the chief masks.

ment which I had always considered fair. I therefore informed him that I had completed a brief manuscript of a scientific nature which I desired to have published. I shall not go into the question here whether he no longer felt that my works would ever again amount to much, or whether, as I suspect in this instance, he had sought professional advice—and this might well be the case with a venture into a new field. Suffice it to say, I found it hard to understand why he refused to print my brochure, when, merely by risking six sheets of paper at the very most, he might have retained for himself a prolific, reliable, easily satisfied author, who was just getting a fresh start.

Once again then I found myself exactly in the same position as when I had offered my *Accomplices* to Fleischer, the bookdealer. However, this time I was not so easily intimidated. Ettinger of Gotha, who had in mind a permanent connection with me, offered to undertake the publication. And so these few pages, elegantly printed in Roman letters, went forth into the world for weal or woe.

The public was taken aback, for inasmuch as it wishes to be served well and uniformly, it demands that every man remain in his own field. This demand is well grounded, for a man who wishes to achieve excellence, which is infinite in its scope, ought not to venture on the varied paths that God and Nature do. For this reason it is expected that a person who has distinguished himself in one field, whose manner and style are generally recognized and esteemed, will not leave his field, much less venture into one entirely unrelated. Should an individual attempt this, no gratitude is shown him; indeed, even when he does his task well, he is given no special praise.

But a man of lively intellect feels that he exists not for the public's sake, but for his own. He does not care to tire himself out and wear himself down by doing the same thing over and over again. Moreover, every energetic man of talent has something universal in him, causing him to cast about here and there and to select his field of activity according to his own desire. We are all acquainted with physicians who are passionate builders, who lay out gardens and factories; with surgeons who are connoisseurs of old coins and possessors of rare collections. Astruc,[58] court physician to Louis XIV, applied probe and scalpel first to the Pentateuch. And to how many other interested amateurs and uninhibited dilettantes is science indebted! In addition we all know businessmen who are passionate cardplayers and readers of novels; we are all acquainted with the serious paterfamilias who prefers the make-believe

---

[58] Jean Astruc, 1684–1766, pioneer in modern Bible research.

of the theater to any other form of entertainment. For the past several years we have heard repeated to the point of irritation the eternal truth that a man's life is composed of both pleasure and work, and that only the man who can maintain a balance between the two deserves to be called wise and happy: for instinctively every person aspires to the exact opposite of his nature, in order that he may achieve a perfect whole.

In a thousand different ways an active man seems impelled to yield to this necessity. Who would remonstrate with Chladni,[59] who is such a great credit to our country! The world is indebted to him for luring sounds in so many ways from bodies of all kinds, for making sound ultimately visible. Yet what could be more remote from such an attempt than his study of atmospheric minerals? It is a splendid pursuit to study and evaluate the details of meteors that keep falling in our own age, to reduce this product of heaven and earth to its constituent parts, and to study the history of this wonderful and ever-recurring phenomenon. But how is the one activity connected with the other? Perchance by the thunderous noise with which these meteoric phenomena plunge down? By no means! It is rather because a man of genius and perception, finding his attention drawn to two widely divergent subjects, now steadily and unswervingly pursues them both. Let us show our gratitude by turning to good advantage the profit we have reaped therefrom.

History of the Printed Brochure[60]

An individual who pursues a worthy subject in private, earnestly endeavoring to master it completely, does not always realize that the viewpoints of his contemporaries may be quite different from his own. This is to his advantage, for he would lose faith in himself if he could not count on sympathetic interest. However, once he comes forward with his opinion he will soon find out that various points of view compete with one another in the world, to the confusion of scholars and ignorant men alike. The times are split by factions that understand themselves as little as they understand their diametrical opposites. All any individual can do is to accomplish enthusiastically whatever he can and succeed as far as possible.

---

[59] Ernst Florens Friedrich Chladni, 1756–1827, German physicist; founder of the science of acoustics, deviser of a mode of exhibiting regions of vibration and rest, since known by his name: "Chladni's figures"; first to express the idea that meteorites were of cosmic origin.

[60] Written in April, 1817; published in *Natural Science in General; Morphology in Particular*, Vol. I, No. 1 (1817).

Even before public criticism reached me, I was altogether dumbfounded over a private communication. In a sizeable German city a group of scientists had founded a society, together accomplishing a great deal of good, both theoretical and practical. In this circle, too, my brochure was eagerly read as a singular novelty. However, all were dissatisfied with it, asserting that there was no knowing what it was intended to mean. One of my Roman artist friends,[61] who loved and trusted me, felt aggrieved to hear my work censured, indeed even condemned, for he had heard me discuss various subjects quite sensibly and logically during long-continued association. He therefore read my brochure with care, and although he did not know exactly what I was driving at, he approached the work with enthusiasm and with an artist's sensitivity, giving my thesis an extremely odd but ingenious interpretation.

"The artist," he said, "has an original and unrevealed purpose, which I nevertheless see through clearly. His purpose is to instruct artists how to represent budding and creeping floral ornaments in the manner of the ancients, that is, as progressive movement. The plant must issue from the simplest leaves, which are gradually diversified, notched, and multiplied. Then by pushing onward, these leaves develop more and more, becoming ever thinner and lighter, until they unite again in the opulence of the flower, either to discharge the seed or perhaps even to begin a new life cycle. Marble pillars ornamented in this fashion are to be seen in the Villa Medici, and only now do I rightly understand what the artist's intention was. The endless abundance of the leaves is finally surpassed by the flowers, so that when at last figures of genii and animals appear in the place of seeds we do not find it in the least strange, following as it does upon a splendid evolutionary process. I am now looking forward to devising a good many ornaments in the manner outlined, for up to now I have been unconsciously imitating the ancients."

In this case, however, a word to the wise was not sufficient. They more or less allowed the explanation to pass, but declared that if the writer had no more than art and embellishments in mind, he should not pretend to be working for the sciences, where such fantasies can have no validity. The artist later assured me that as a result of my formulation of natural laws he had succeeded in combining the natural and the impossible to produce a result agreeably probable. However, he did not dare approach those gentlemen again with elucidations.

From other sources I heard similar tales. Nowhere would anyone

---

[61] According to Goethe's secretary Friedrich Riemer, the friend was Johann Heinrich Wilhelm Tischbein, 1751–1829, historical painter, whose famous portrait of Goethe in Italy is reproduced as the frontispiece of this volume.

grant that science and poetry can be united. People forgot that science had developed from poetry and they failed to take into consideration that a swing of the pendulum might beneficently reunite the two, at a higher level and to mutual advantage.

My women friends, who formerly had wanted to take me away from lonely mountains and from my study of lifeless stones, again were far from satisfied with my abstract gardening. In their opinion, flowers and plants should have form, color, and fragrance, whereas in my work they faded out into wraithlike figures. And so I attempted to entice the interest of these well-disposed ladies by an elegy, which I should like to insert here in the context of a scientific treatise, where it might possibly be better understood than if it were inserted in a series of tender and impassioned love poems.

> The rich profusion thee confounds, my love,
> Of flowers, spread athwart the garden. Aye,
> Name upon name assails thy ears, and each
> More barbarous-sounding than the one before—
> Like unto each the form, yet none alike;
> And so the choir hints a secret law,
> A sacred mystery. Ah, love could I vouchsafe
> In sweet felicity a simple answer!
> Gaze on them as they grow, see how the plant
> Burgeons by stages into flower and fruit,
> Bursts from the seed so soon as fertile earth
> Sends it to life from her sweet bosom, and
> Commends the unfolding of the delicate leaf
> To the sacred goad of ever-moving light!
> Asleep within the seed the power lies,
> Foreshadowed pattern, folded in the shell,
> Root, leaf, and germ, pale and half-formed.
> The nub of tranquil life, kept safe and dry,
> Swells upward, trusting to the gentle dew,
> Soaring apace from out the enfolding night.
> Artless the shape that first bursts into light—
> The plant-child, like unto the human kind—
> Sends forth its rising shoot that gathers limb
> To limb, itself repeating, recreating,
> In infinite variety; 'tis plain
> To see, each leaf elaborates the last—
> Serrated margins, scalloped fingers, spikes
> That rested, webbed, within the nether organ—
> At length attaining preordained fulfillment.

ON HIS PLANT STUDIES

Oft the beholder marvels at the wealth
Of shape and structure shown in succulent surface—
The infinite freedom of the growing leaf.
Yet nature bids a halt; her mighty hands,
Gently directing even higher perfection,
Narrow the vessels, moderate the sap;
And soon the form exhibits subtle change.
The spreading fringes quietly withdraw,
Letting the leafless stalk rise up alone.
More delicate the stem that carries now
A wondrous growth. Enchanted is the eye.
In careful number or in wild profusion
Lesser leaf brethren circle here the core.
The crowded guardian chalice clasps the stem,
Soon to release the blazing topmost crown.
So nature glories in her highest growth,
Showing her endless forms in orderly array.
None but must marvel as the blossom stirs
Above the slender framework of its leaves.
Yet is this splendor but the heralding
Of new creation, as the many-hued petals
Now feel God's hand and swiftly shrink. Twin forms
Spring forth, most delicate, destined for union.
In intimacy they stand, the tender pairs,
Displayed about the consecrated altar,
While Hymen hovers above. A swooning scent
Pervades the air, its savor carrying life.
Deep in the bosom of the swelling fruit
A germ begins to burgeon here and there,
As nature welds her ring of ageless power,
Joining another cycle to the last,
Flinging the chain unto the end of time—
The whole reflected in each separate part.
Turn now thine eyes again, love, to the teeming
Profusion. See its bafflement dispelled.
Each plant thee heralds now the iron laws.
In rising voices hear the flowers declaim;
And, once deciphered, the eternal law
Opens to thee, no matter what the guise—
Slow caterpillar or quick butterfly,
Let man himself the ordained image alter!
Ah, think thou also how from sweet acquaintance
The power of friendship grew within our hearts,
To ripen at long last to fruitful love!

> Think how our tender sentiments, unfolding,
> Took now this form, now that, in swift succession!
> Rejoice the light of day! Love, sanctified,
> Strives for the highest fruit—to look at life
> In the same light, that lovers may together
> In harmony seek out the higher world!
>
> [Translated by Heinz Norden in Rudolf Magnus, *Goethe as a Scientist* (Henry Schuman, 1948)].

This poem was most welcome to the beloved person who had the right to believe that the charming figures of speech had reference to her.[62] And I too was very happy to see that the lovely parable increased and perfected our great attachment. But from the other charming ladies I had much to suffer, for they wrote parodies of my metamorphoses, with droll and bantering allusions.

Sorrows of a more serious nature came to me on the part of friends outside my immediate circle to whom I had jubilantly distributed my free copies. But they all answered more or less in the idiom of Bonnet,[63] for his *Contemplation de la nature* had won intellects by its deceptive lucidity and had initiated a language in which they thought they were actually saying something and being understood. No one cared to accommodate himself to my method of expressing myself. It is most tormenting not to be understood when one feels sure himself, after great stress and strain, that one understands both one's self and one's subject. It drives one to insanity to hear repeated again and again a mistake from which one has himself just escaped by a hair's breadth, and nothing is more painful than to have the things that should unite us with informed and intelligent men give rise instead to unbridgeable separation.

Moreover, my friends were by no means tactful in their comments, and author of long standing though I was, I had occasion to learn that it is precisely through presentation copies that one may experience vexation and disgust. If an individual gets hold of a book by accident, or because someone recommends it to him, he reads it and may then even purchase a copy for himself; but if his friend, the author, with happy expectancy presents the opus to him, it appears as if the author were attempting to weight the intellectual scales in his own favor. Then one of the cardinal sins emerges in its most loathsome form, envy and ill will, just as with jubilant lovers when they are confiding to friends an affair of the heart. Several friends whom I consulted were likewise

---

[62] Goethe's wife, Christiane Vulpius. See p. 251.
[63] Charles de Bonnet, 1720–1793, exponent of the pre-formation theory, the more extreme version of which stood irreconcilably opposed to Goethe's theory.

not unacquainted with this phenomenon of the civilized world.

At this point I must commend one friend who did aid me faithfully, both during and after completion of my study. It was Karl von Dalberg,[64] a man who indeed was deserving of a more peaceful world in which to attain the good fortune to which he was born and destined, in which to adorn the highest positions through untiring activity, to enjoy with his loved ones the advantages accruing therefrom. Friends always found him alert, interested, and helpful; even though they could not always make use of his point of view in its entirety, they nevertheless found him intelligent and helpful in individual matters. Throughout my scientific work, I have been greatly in debt to him; he had the knack of stimulating and enlivening my characteristic method of observing Nature; and he had the boldness to catch hold of what we were studying with decisive convenient phrases, thus bringing it within the range of understanding.

A favorable critique in the *Göttingen Review* could satisfy me but partly. It was admitted that I had treated my subject with unusual clarity; the development of my thesis was presented briefly and clearly by the reviewer; but its significance was not touched upon and for that reason I did not benefit by the review. But since it was admitted that I had beaten a path to knowledge, I fervently hoped that I would be met halfway: for with me it was not a question of merely getting a foothold, but rather of striding through these regions as soon as possible, instructed and enlightened. Since things did not turn out as I had hoped and wanted, I held to my previous arrangements. Herbaria were collected for this purpose; I even preserved some oddities in alcohol; I had drawings and illustrations prepared, all of which were to benefit my work.

My purpose was to visualize the chief phenomenon and to prove my theory through application. At this point, however, I was unexpectedly caught up in a life of stirring activity—I went with my prince and the Prussian army to Silesia,[65] to Champagne,[66] to the siege of Mayence.[67] Yet the following three years were also of advantage to my scientific aspirations. I observed natural phenomena out of doors and thus found it wholly unnecessary to allow a hair's breadth of sunlight to enter a dark room to demonstrate that light and darkness produce colors.[68] In that way I scarcely noticed the endless boredom, which is

---

[64] Karl Theodor von Dalberg, 1744–1817, emissary at Erfurt of the Electoral Prince Bishop of Mayence, which latter office he himself assumed in 1802.
[65] 1790.   [66] 1792.   [67] 1793.
[68] An allusion to the Newtonian experiment for the theory of color, which was so abhorrent to Goethe.

as deadening as danger is exhilarating and diverting. My observations went on without interruption and were written down without intermission. The guardian angel who had previously stood by my side at Karlsbad was here again standing at my side to goad me on.

Because all opportunity of consulting books was impossible for me at that time, I occasionally used my brochure in approaching learned friends who were interested in the subject, with the request that they kindly be alert in their wide reading to what had already been published on the subject; for I had long been convinced that there was nothing new under the sun and that one can find hints in transmitted knowledge of what we ourselves are just discovering, thinking about, and even producing. We are original only because we know so little.

That wish was very happily fulfilled when my honored friend, Friedrich August Wolf,[69] called my attention to a namesake of his who had long been following the trail that I was now pursuing. How I profited by the information we shall now see.

My Discovery of a Worthy Forerunner[70]

CASPAR FRIEDRICH WOLFF[71] was born in Berlin in 1733, studied at Halle, receiving his degree in 1759. His dissertation, *Theoria generationis,* exhibited a range of microscopic investigation and serious sustained thought quite rare in a young man of 26. Next, he practiced in Breslau, at the same time lecturing on physiology and other subjects in the local hospital. Called to Berlin, he continued lecturing; and wishing to give his listeners a complete presentation of his generic theory, he published in German in 1764 an octavo, of which the first part is historic and polemic, and the second, systematic and didactic. Hereupon he was transferred to St. Petersburg as a professor, where, if one may judge from his commentaries and reports from 1767 to 1792, he was an industrious collaborator. His essays all prove that he remained absolutely true to his plan of study, and to his convictions as well, until his death in 1794. His colleagues had the following to say about him:

"He had brought with him to St. Petersburg a well-established reputation as a thorough anatomist and a profound physiologist, a reputa-

---

[69] 1759–1824; German classical scholar and critic.

[70] This was written during the period October, 1816, to May, 1817, and was printed in *Natural Science in General; Morphology in Particular,* Vol. I, No. 1 (1817).

[71] Goethe misspelled the name, thus: Wolf.

tion which he preserved and increased by numerous and excellent articles in the publications of the Academy. Earlier he had acquired fame through a solid and deeply conceived preliminary report on generation and his ensuing dispute with the immortal Haller,[72] by whom he was always treated with respect and goodwill in spite of their difference of opinion. Loved and honored by his colleagues for his gentleness and his integrity as well as for his knowledge, he died in the sixty-first year of his life, mourned by all members of the academy in which he had been so active a member for twenty-seven years. Neither the family nor the papers of the deceased could supply details from which to compile a detailed biography. However, the unvarying schedule of this scholarly recluse, who spent his whole life in his library, was unlikely to yield much biographical material, so that we are probably not missing much. After all, it is in his writings that the essential, the significant, the useful part of such a man's life is preserved, and it is through them that his name is handed down to posterity. Thus in the absence of biographical material we are presenting a register of his academic works, for such a list may indeed serve as a eulogy and will make us feel the extent of our loss more than the most beautiful turns of speech could do."

In this manner, as long as twenty-five years ago, a foreign nation paid homage to this admirable compatriot of ours, a man driven from his homeland early in his career by a reigning school of thought to which he could not subscribe. And I am happy to confess that for more than twenty-five years I have been learning from and through him. How little known Wolff was during this time in Germany, was told by Meckel,[73] as honest as he is deserving, on the occasion of his translation of Wolff's essay, *On the formation of the intestinal canal in the hatched chick* (Halle, 1812).

May the Goddess of Fate grant me the power to relate in detail how for many years I walked at the side of this distinguished man; how I sought to understand his character, convictions, and theories; the extent to which I could agree with him; and my urge to progress beyond him, though all the while keeping him gratefully in view. At present we are concerned only with his theory on plant formation, a theory outlined in his preliminary report and in an extended German version of it, but summarized and expressed most clearly in the essay listed first in the Academy's register of his works. Therefore, I gratefully present the

---

[72] Albrecht von Haller, 1708–1777, Swiss physiologist, anatomist, and botanist, exponent of the theory of pre-formation.

[73] Johann Friedrich Meckel, 1781–1833, anatomist, the youngest and most famous of his name. Wolff's essay had appeared in 1768–1769.

following passages from Meckel's translation, merely appending a few comments to indicate the points I shall develop later in greater detail.

Caspar Friedrich Wolff on Plant Formation

"I attempted to explain most of the plant parts, those which have the greatest similarity with one another and on that account are easily compared,—namely, calyx, leaves, fruit hull, seed, stem, root—according to their genesis. In doing so, I confirmed the fact that the various parts comprising the plants are extraordinarily similar to one another and for that reason can easily be recognized from their nature and their manner of origin. Actually, it does not require a great deal of acuteness to notice that the calyx is only slightly different from the leaves and, to put it briefly, is nothing more than a collection of several smaller and less developed leaves. This is seen very clearly in several annual plants with composite leaves, in which the leaves gradually become smaller, less developed, and more numerous, the higher they stand on the stem, until finally the last ones, which are directly below the flower and are extremely small and closely packed together, become the leaves of the calyx and taken together form the calyx itself.

"It is just as clear that the seed pod too is made up of several leaves, except that in this instance they are fused with one another, whereas in the calyx they were merely crowded together. The correctness of this view is proved not only by the way in which a number of seed pods open up and divide, by themselves, into the leaves which form their component parts, but also by mere examination and the outward appearance of the pod. Finally, even the seeds themselves are really nothing more than leaves, notwithstanding the fact that at first glance they have not the slightest similarity to them; for the lobules into which they are split are leaves; however, they are the least developed leaves of the whole plant, misshapen, small, thick, hard, sapless, and white. Any doubt concerning the truth of this assertion is dispelled on observing that when the seed is put into the earth to continue the growth interrupted in the mother plant, these lobules are instantly transformed into the most perfect, green, juicy leaves, the so-called seed leaves. Moreover, from isolated cases it appears at least possible that the corolla and stamens are nothing more than modified leaves. For it is no rarity to see the leaves of the calyx transformed into petals, and conversely to see the petals transformed into sepals. But if the sepals are true leaves and the petals nothing more than sepals, then the petals too are undoubtedly modified genuine true leaves. Similarly, one observes that the stamens in the Linnaean Polyandria are frequently transformed into petals, thereby creating double flowers, and conversely that the petals are transformed into stamens; from this fact it may be concluded that the stamens, too, are essentially leaves. In a word, mature reflection reveals that the plant, the various parts of which appear so extraordinarily different from one another at first glance, is composed exclusively of leaves and stem, inasmuch as the root is part of the stem. The leaves and stem are the nearest, the most direct and complex plant parts; the veins and vesicles are the remote and simple parts, from which in turn the first are formed.

"If all plant parts with the exception of the stem can be derived from the leaf form and are nothing more than modifications of it, it follows that it would not be hard to evolve a generation theory of plants; at the same time the path is indicated which one must take to prove the theory. First, one must discover through observation the way in which the leaves proper are formed, or, what amounts to the same thing, the way normal vegetation takes place, the basis on which it rests, and the forces whereby it is effected. After this has been determined, we must investigate—in the upper part of the plant, where seemingly new phenomena and seemingly different parts develop—the causes, circumstances, and conditions which modify the general manner of vegetation to the extent that in place of the usual leaves these oddly formed ones make their appearance. Proceeding in accordance with this plan, I found[74] that all these modifications have their basis in the gradual decrease in the vegetative power, which diminishes in proportion to the duration of vegetation and finally disappears completely; and that consequently the essence of all these changes of the leaves is their less perfect development. By numerous experiments I found it easy to demonstrate this gradual diminution of vegetation and its cause—an exact account would be too long to insert at this point—and to explain solely on this basis the new phenomena exhibited by the flower and fruit parts, apparently so different from the other leaves, and even to explain numerous small related details.

"This is how matters stand when it is the history of plant formation that is being studied; everything is quite different, however, when one turns to animal life."

---

[74] Here Wolff's theory of metamorphosis differs decidedly from Goethe's, for where Goethe sees the growth increasing to an ennobled form, i.e., sexual propagation, Wolff assumes a decrease in the vegetative power.

A Few Additional Comments

In appending a few remarks to the preceding passages I must avoid entering too deeply into an explanation of this excellent man's views and theories. Such an attempt I may possibly make at some later time; here I am presenting merely enough to stimulate further thought.

Wolff explicitly recognizes the identity of plant parts in spite of their variability; but his experimental method, once he has adopted it, prevents him from taking the final and decisive step. That is to say, because the theory of pre-formation and insertion which he opposes, rests on a mere speculative concept—in other words, rests on an assumption that seems plausible but cannot be made evident to the senses—he sets forth as a basic maxim of all his own experiments: that one may assume, grant, and assert nothing except what one has seen with one's own eyes and what one is at all times in a position to produce again. For this reason he is constantly endeavoring to explore the origins of life through microscopic studies, and in that way to follow organic embryos from earliest appearance to maturity. Excellent as this method is, and however much he may have accomplished with it, the worthy man nevertheless failed to realize that there is a difference between seeing and seeing; he failed to realize that the intellectual eye must work in constant and spirited harmony with the bodily eye, for otherwise the scholar might run the risk of looking and yet overlooking.

In the transformation of plants he saw the same organ always contracting, getting smaller. The fact that this contraction alternates with an expansion, he did not see. He saw that it decreased in volume without noticing that it was at the same time perfecting itself, and he absurdly attributed to degeneration this path toward perfection.

In that way, he himself closed off the path by which he might attain directly to the metamorphosis of animals; indeed, he takes an explicit stand against this latter concept; it was an entirely different thing with the metamorphosis of animals, he said. However, since his investigational method was the correct one and his power of observation most exact, and since he insisted that detailed study and tracing of organic development must antecede any description of the finished part, he always arrived at the correct result even when he was in contradiction with himself.

Therefore, if in one place he denies analogy of form in the various organic parts of the animal's interior, in another place he is eager to grant it. The fact that he has compared certain individual organs that actually have nothing in common—for example, intestinal canal with liver; liver with brain—leads him to the denial of analogy in form. On the other hand, he comes to an affirmative conclusion when he compares system with system, since the analogy then strikes his eye immediately, and he soars to the bold idea that there might be an assemblage of several animals here.

At this point I can stop with a good conscience, now that one of his most outstanding works has been made available to Germans through the services of our esteemed Meckel.

THREE FAVORABLE REVIEWS[75]

AUTHORSHIP IS CERTAINLY an odd thing! To make too little or too much of one's achievement is equally wrong. However, a man of spirit naturally wishes to exert an influence upon his time and therefore hopes that his contemporaries will not keep silent about him. With regard to aesthetic works I have had no occasion to complain about my contemporaries; yet here I had formed my own independent opinion, thus deriving from approbation little enjoyment and from disapproval little chagrin. Youthful lightheartedness, pride, and arrogance helped me over all situations that might have been somewhat unpleasant. Then, too, on a higher plane, the realization that one is alone, and must do everything by himself, that no one can help in these creative productions, gives to the spirit such power as to make one feel superior to all obstacles. Also, it is a gracious gift of Nature that the art of creation is itself a pleasure and its own reward, so that one does not feel justified in demanding anything more.

In science, I have found the situation different. To attain any sort of footing and firm ground in this field requires industry, toil, and exertion. What is even more, we feel that here the individual is not self-sufficient. A glance at history suffices to show that it has required a succession of gifted men, over a period of centuries, to win from Nature and mankind what little we know. From year to year we see new discoveries being made and become convinced that the field of science is unlimited.

Here, then, we work earnestly not on our own behalf, but on behalf of a worthy cause; and we ourselves demand recognition in the same way

---

[75] Only two reviews are mentioned in the following comments, written in September, 1817, published in *Natural Science in General; Morphology in Particular*, Vol. I, No. 2 (1820). Undoubtedly Goethe had in mind, as the third, the one in the *Göttingen Review* of 1791 (see p. 175).

that we recognize the efforts of others. We long for assistance, sympathy, and encouragement. In my case there would have been no lack of these either, had I been more alert to what was going on in the world of scholarship. However, my restless efforts to develop all sides of my personality, coming just at the moment when world-shaking events were disturbing us from within and assailing us from without, made it impossible for me to learn what the world was thinking about my scientific works. This explains my peculiar experience in not getting to see until very late two extremely favorable reviews of *The Metamorphosis of Plants:* one in the *Gotha Scholarly Review* of April 23, 1791, and the other in the *Universal German Library,* Vol. CXVI, p. 477; and, quite as though a propitious destiny had intended to keep something pleasant in reserve for me, they came to my attention just at a time when, in a different field, I was being most viciously attacked from all quarters.

Other Friendly Overtures[76]

IN ADDITION to this encouragement, I was gratified by having my little essay included in a Gotha encyclopaedia, an indication to me that my work was at least considered as having some value for the general public.

Jussieu[77] had mentioned the *Metamorphosis* in his introduction to plant theory, but merely in the passage on double and anomolous flowers. Moreover, he did not make clear that the law of regular formation might also be operative here.

Usteri,[78] in his supplement to the introduction of the Zürich edition of Jussieu's book in 1791, promised to express an opinion on this subject, thus: *De Metamorphosi plantarum egregie nuper Goethe V. Cl. egit, ejus libri analysin uberiorem dabo.*[79] Unfortunately, the ensuing stormy events deprived us, and me most of all, of the comments of this outstanding man.

Willdenow,[80] in his outline of botany in 1792, gives no recognition to my work. However, it is not unknown to him, for on page 343 he says, "The life of the plant is, therefore, as Herr Goethe very nicely puts it, an expansion and contraction, and these alternations comprise the

---

[76] Written in 1817 and first printed in *Natural Science in General; Morphology in Particular,* Vol. I, No. 2 (1820).

[77] Antoine Laurent de Jussieu, 1748–1836, French botanist.

[78] Paulus Usteri, 1768–1831, director of the botanical garden in Zürich.

[79] "The famous Goethe has recently written an excellent essay on plant metamorphosis. I shall later return to this publication in greater detail."

[80] Karl Ludwig Willdenow, 1765–1812, well-known Berlin botanist.

various stages in its life." I can let the "nicely" pass, especially in view of the august source of the citation; but the *"egregie"* of Herr Usteri is nevertheless "nicer" and more complimentary.

Other scientists also accorded me some attention. Batsch,[81] in token of his sympathy and gratitude, coined the term *Goethia* and was good enough to place it under the Sempervivum; however, it did not maintain its place in the system. What its name may be at present, I am in not a position to say.

Kindly disposed men in the Westerwald region, discovering a beautiful mineral, called it "goethite" as a favor and honor to me. I still owe a great deal of thanks to Messrs. Cramer[82] and Achenbach[83] for that, although this term also disappeared rapidly from oryctognosy.[84] It was also called ruby blende, and at present it is known by the term Pyrosiderite.[85] It was enough for me that in connection with such a beautiful product of Nature one should have thought of me even for a moment.

A third attempt to erect a monument to my name in science was recently made by Professor Fischer,[86] in memory of former good fellowship. In 1811 he published in Moscow *Prodromum craniologiae comparatae*,[87] in which he lists *Observata quaedam de osse epactali, sive Goethiano palmigradorum*[88] and pays me the honor of bestowing my name on that division of the occiput to which I had given some consideration in my investigations. But without doubt this good intention will fail of its purpose too, and I shall have to resign myself, as in other cases, to seeing this friendly memorial disappear from scientific terminology.

Should my vanity be at all offended at the thought that I am to be remembered neither by the names of plants and minerals, nor of ossicles, I can nevertheless take consolation in the benevolent interest of a highly esteemed friend. With the German translation of his ideas on a combined geography of plants and a nature description of the tropics, Alexander von Humboldt[89] sends me a flattering picture in which he

---

[81] See pp. 76, 96, 155, 189, 196, 217.
[82] Ludwig Wilhelm Cramer, mining councillor in Wiesbaden, Goethe's friend.
[83] Heinrich Adolf Achenbach, rector at Siegen.
[84] Descriptive mineralogy.
[85] The correct form is "Pyrrhosiderite."
[86] Gotthelf Fischer von Waldheim, 1771–1853, professor of natural history, first in Mayence and later in Moscow.
[87] "Introduction to Comparative Craniology."
[88] "Observations on the epactal or Goethe bone of the palmegrades."
[89] In 1807 Humboldt dedicated his "Ideas on a Geography of Plants" to Goethe. The dedication plate, drawn by the Danish sculptor, Thorvaldsen, represented Apollo lifting the veil of the goddess of Nature. At the base of the goddess was an inscription, "Metamorphosis of Plants." Under the picture as a whole stood the words, "To Goethe."

suggests that poesie, too, can succeed in lifting the veil of Nature. And if *he* admits this, who would dare deny it? I therefore feel an obligation to express my thanks publicly.

Perhaps it would also be suitable to give grateful acknowledgment here to several academies of science and other organizations active in furthering science which have expressed the desire to include me as a member. Lest anyone find fault with me for giving all this information about myself so outspokenly, lest anyone regard that sort of thing as unseemly praise, I shall soon take the opportunity of speaking just as freely and unreservedly about unfriendly and offensive treatment of my scientific efforts in a related field over a period of twenty-six years.

But now let us turn to other pleasant endeavors in the bright world of plants, especially since another highly gratifying reward for my labors and perseverance reaches me just as I am about to send these pages to press. Perusing our worthy Kurt Sprengel's[90] history of botany, for the purpose of surveying this estimable science, I find my own work favorably recognized. Can there be a greater reward than winning approval from the very men whom one has held up as ideals all during one's undertaking?

It is our good fortune if, as we advance in years, we have no reason to complain of the changing character of the time. Youth longs for sympathetic interest, maturity demands recognition, and old age hopes for affirmation. Although youth and maturity usually receive their allotted portion, old age is often cheated of its reward. Even if an old man does not outlive himself, others nevertheless survive him, and advance beyond him; modes of thought and experimentation of which he never dreamed take hold and spread.

My own lot has been a desirable one. Young men have found their way to the road which I have travelled, partly upon the trail which I myself had blazed and partly upon the trail opened by the spirit of the times. From now on, hesitation and hindrance are almost unthinkable; indeed, overhaste and exaggeration are probably more likely than retrogression and stagnation. In these good days, enjoyed and appreciated by me, it is hard to recall that circumscribed period when no one would come to the aid of an earnest, faithful endeavor. A few incidents may be recalled here as examples.

Hardly had my first little essay devoted to Nature created a sensation, albeit an unfavorable one, when I met upon my journies a worthy, aged man whom I came to venerate in every sense of the word and—because he continued to encourage me—to love. After welcoming me cheerfully,

---

[90] Kurt Sprengel, 1766–1833, professor of botany at Halle.

he remarked somewhat dubiously that he had heard that I had just begun a study of botany. From such a course he had every reason to earnestly dissuade me, for his own attempt to approach this branch of science had miscarried. Instead of joyous Nature he had discovered nomenclature, terminology, and a pedantry so timid that it killed the spirit, hindered and paralyzed all free movement of the intellect. With the best of intentions, he was therefore advising me not to exchange the ever-blooming fields of poesy for provincial flora, botanical gardens, and greenhouses, not to mention dried-out herbaria.

Although immediately foreseeing how hard it would be to enlighten my well-intentioned friend about my end-purpose and my labors, and to convince him, nevertheless I confessed that I had just published a brochure on the metamorphosis of plants. Without allowing me to finish he joyfully declared that now he was satisfied, consoled, and cured of his mistake. He realized indeed that I had treated my theme in the Ovidian style[91] and that he was already looking forward to discovering the charming way in which I had portrayed the hyacinth, the sweetpea, and the narcissus. Then the conversation turned to subjects that had his complete approval.

With such resolution were my hopes and desires denied, lying as they did entirely outside the intellectual horizon of the time. Each activity was considered apart from the others; science, the arts, commerce, industry, and everything else imaginable all moved in their own isolated orbits. The individual scientific worker was of course in earnest, but for that reason worked only for himself and in his own way; his neighbor remained a complete stranger, and the two alienated each other. Art and poetry scarcely ever coincided, and lively interaction of the two was out of the question; poetry and science seemed, indeed, to be the greatest adversaries. Thus, because the individual spheres of operation were detached from one another, the operations within each sphere were also isolated and dispersed. Even so much as a hint of theory created terror, because for more than a century scientists had been avoiding theory as though it were a spectre, though in the end, as was inevitable with such fragmentary experience, they had embraced the most vulgar conceptions. No one would admit that an idea or concept is the basis for all observing—an idea or concept which will foster experimentation, indeed promote discovery and invention.

It was bound to happen, however, that somewhere in writing or in

---

[91] In Ovid's *Metamorphoses* the transformation of human beings into animals and plants is described in the form of myths; Hyacinthus, Clytia, and Narcissus are some of the mythical figures who are transformed into beautiful flowers.

conversation, I should make a remark so pleasing to men of that sort, that they were willing to receive and adopt it as an isolated observation. In such cases they extolled, they called it a stroke of luck, crediting the person who had made the observation with a degree of discernment, for in individual details they themselves possessed discernment. Thus, by giving so much credit to anyone else for an isolated observation apart from its context, they simply provided one more example of their inconsistency.

Notes for an Essay on Plant Culture in the Grand Duchy of Weimar[92]

This highly significant and striking achievement also sprang from genuine activity, from serene and cheerful co-operation, successfully maintained over a number of years.

Let us first discuss Belvedere, flourishing and thriving, the joy of our native Weimarians, the admiration of our visitors.

The castle and garden were completed in 1730 by Duke Ernst August,[93] and set aside as a pleasure resort for the Ducal Court. The hillside forests in the background were rendered charming and accessible through promenades, pavilions, and other romantic structures. A large orangery was built—and whatever else that was considered an absolute necessity for such gardens at that time; and alongside the orangery was a small menagerie consisting chiefly of exotic birds. In this manner gardening and horticulture were carried on and fostered. Nor was the cultivation of a number of kitchen plants neglected.

However, it is impossible to describe the normal course of events and the steadily expanding cultivation of plants—of interest to botanists as well as to lovers of landscape gardening—without first telling about the castle fire.

Members of the high nobility, deprived as they were by the fire of their comfortable residence, and lodged temporarily in quarters totally unsuited to people of their high rank, turned to the out-of-doors. The various well-arranged summer houses, and especially the pleasant Ilm valley near Weimar with its relatively old flower and vegetable gardens, provided an enchanting opportunity.

---

[92] Written during the period 1820–1822; printed in *Natural Science in General; Morphology in Particular*, Vol. I, No. 4 (1822). As the title indicates, the article is in the form of rough notes.

[93] Ernst August, 1688–1748, duke of Saxe-Weimar.

The park in Dessau, among the first and best to become famous and to attract tourists, inspired them to create something similar. This emulation could take a most original form since the two localities did not resemble each other in the least. The flat open landscape of Dessau, with its numerous rivers and lakes, had nothing in common with that of Weimar, which was alternately hilly and flat. The peculiar charm of the latter was successfully enhanced; the friendship of the two princes, who frequently exchanged visits, and especially Hirschfeld's[94] love of landscape gardening, provided the opportunity to determine by comparison what was most suited to each region.

The appointment of Court Gardener Reichert to Belvedere very soon created an opportunity to satisfy all such wishes. An expert in large-scale propagation, he not only carried on such work in Belvedere but soon laid out a commercial garden of his own in the vicinity of Weimar. Bush and tree plantings were therefore propagated every spring and autumn.

With the improvement of the region, the desire to enjoy life in the open has increased. Little houses have been built which are suitable for visitors to the country and which at least do not mar the beauty of the landscape, and perhaps actually enhance it. They give opportunity for comfortable accommodation of large and small gatherings, in fact provide impetus for parties in the open; for such events the diversified terrain offers great variety and favors many a surprise, since vivid imagination and inventiveness of combined talents can find expression in many ways.

Thus the park grounds increase in size, extending directly from the castle—which is becoming habitable again as it gradually rises from its ruins—up along the Ilm valley almost to Belvedere. The prince himself undertakes chief supervision, management, and direction, while Her Highness, the Ducal Consort, also participates, through her constant interest and painstaking cultivation of plants.

The Duchess Amalia's[95] sojourn in Ettersburg and Tiefurt contributes not a little to what might almost be termed a passionately felt need for outdoor life.

At the last-mentioned place Prince Konstantin[96] and Major von Knebel[97] had already done preparatory work for many years, and had inaugurated this most pleasant Ilm valley as a place for gay parties and outings.

---

[94] Christian Kajetan Lorenz Hirschfeld, 1742–1792, professor of aesthetics, Kiel; author of a book on the theory of landscape gardening.
[95] Anna Amalia, 1739–1807, dowager duchess of Saxe-Weimar, mother of Karl August.
[96] Son of Karl August.
[97] Karl L. von Knebel, 1744–1834, tutor of Prince Konstantin. See p. 3.

In general, an effort is being made at all times to do justice to the locality, to utilize it to the fullest extent, and to make no provisions that are inconsonant with its basic character.

The regular cultivation of forests assiduously goes forward, and with it the nurture of exotic tree varieties.

Large plantings and other forms of propagation are being conducted by intelligent foresters, and information is being gained thereby as to which plants will withstand our climate.

At this point some details should be inserted on the somewhat inclement climate of Weimar and Belvedere. Our elevation above sea level is considerable; another consideration is the presence of the Thuringian Forest located to the south of us. Not infrequently our vegetation is menaced by the northeasterly and northwesterly prevailing winds.

Reichert, ducal court gardener, has died; by this time the botanical cultivation of park plants is extensive. Of the stock, a large portion falls to the share of the Prince; agreement is being reached regarding another portion; Reichert's son will take the remainder to Weimar as his own share.

Efforts of other private individuals should be mentioned, especially those of Bertuch,[98] Councillor of the Legation, who has managed the park grounds for sixteen years under the direction of His Serene Highness, besides cultivating a considerable garden of his own and writing various monographs.

Sckell,[99] Inspector of Gardens, has been appointed to Belvedere. He and his brother manage the institution well and faithfully. The son of the former is sent on a journey; as are several other sons also. They come back after a while with important plant shipments.

The botanical garden proper continues to be directed by His Serene Highness, the Grand Duke. The castle and remaining country residences have been made available to the ducal family.

The acquisition of rare botanical works for the public library goes forward. Indeed, the library is increasing by leaps and bounds.

In the same enthusiastic way propagation of important plants and the continued arrival of exotic plants make imperative the extension of Belvedere along the hillside as well as into the valley to the south. In the latter area, greenhouses designed by Their Royal Highnesses are being constructed; of late a palmery of striking appearance had been erected.

---

[98] Friedrich Johann Justin Bertuch, 1747–1822, bookseller, author, officer of the Weimar Court.

[99] Johann and Christian Sckell (or Skell), of a famous family of horticulturists. See p. 150.

ON HIS PLANT STUDIES   189

Buildings in which exotic plants can remain in the earth and merely be covered over in winter, so-called conservatories, have long since been erected and are now being enlarged.

It has been decided to reserve the botanical gardens at Belvedere for scientific purposes exclusively. Therefore, the vegetable garden, the cultivation of pineapples, and similar projects are being transferred to the park in Weimar.

The travels to France, Holland, and Lombardy of Their Serene Highnesses,[100] their visits to botanical gardens, their personal studies of various institutions, all contained in written reports, are instructive and helpful in stimulating interest; in the same way their personal acquaintance with professional scientists and artists has always proved beneficial. Their Highnesses are elected to the London Horticultural Society as regular members of the first rank.

What took place in Jena should not be overlooked. Years ago our good Batsch[101] had already planted part of the ducal garden in accordance with the family system.[102] This arrangement was faithfully carried forward by Schelver[103] and Voigt.[104] The latter reworked the Belvedere as well as the Jena catalogue in accordance with the system mentioned. Yet from time to time, when making direct sales or exchange, the serviceable method used previously was re-employed.[105]

Meanwhile, the expansion of the Belvedere establishment continues without interruption. At this time it becomes apparent that considerable difference of opinion prevails with respect to nomenclature, classification of plants and species, indeed of varieties, this difference of opinion being revived from time to time by visiting connoisseurs and scientists.

Meanwhile, the need becomes more and more imperative for a purely scientific catalogue, as a precise and reliable reference for personal use and for use in sales and exchanges. This lengthy undertaking, if it is to be handled conscientiously, properly requires the appointment of a scientist. For this work Professor Dennstedt[106] is commissioned; he undertakes the work; the first number of the catalogue appears in 1820, the second in 1821. In this way not only the purposes mentioned above are being taken care of, but in addition many an uncertain and inexperienced

---

[100] Goethe facetiously and affectionately uses the Latin *Serenissimi* in referring to his friends, the Duke and Duchess.
[101] See pp. 76, 96, 155, 183, 196, 217 for other references to Batsch.
[102] That is, according to Jussieu's classification by natural families.
[103] See pp. 105, 106, 119, 199, 241 for other references to Schelver.
[104] See pp. 121, 122, 198, 201, 206 for other references to Voigt.
[105] That is, the Linnaean system.
[106] August Wilhelm Dennstedt, professor of botany in Belvedere.

gardener is being provided with a guide for acquiring more detailed botanical knowledge.

Moreover, an extraordinary service might have been rendered to science by the catalogue if diacritical marks had been used with the names and if accents had been indicated occasionally. For at present, from within and from without, from natives and from visitors, we listen to a Babylonian confusion, not of tongues but of pronunciations, arousing in the midst of these magnificent natural creations irritation and ill-humor, especially on the part of individuals familiar with Greek derivatives.

In keeping with the liberality of His Serene Highness and with his truly princely passion for sharing everything good and useful, Jena benefited in proportion to Belvedere's growth. Numerous gifts of plants and seeds from Belvedere were eagerly planted in Jena's new conservatory—a structure seventy-four feet in length, with several sections, built with information based on the latest experiments. The new building itself cut off a large part of the earlier projects of Batsch and in consequence the hotbeds also had to be transferred, making a complete replanting of the whole garden necessary. In connection with this, an improved regrouping of the plant families was made desirable and possible.

We may indeed confidently invite all plant lovers to Belvedere, and to Jena as well, but wish that we were in a position to furnish them with a more exact guide.

Before closing we ought to touch on a good many other things, but we shall mention only the large tree nursery of prolific stock which has been in existence for many years, under the supervision of Legation Councillor Bertuch. Unfortunately, we are losing this man, indefatigable in this as in other fields, at the very moment when we are completing the sketch of this essay, with which—because of his remarkable memory—he might have rendered us most valuable assistance, both with the details and with the general point of view. I might justly deserve to be reproached for my irresponsible procrastination during our long and happy association if it were not for the fact that every individual's life is so teeming with activity that his immediate interest tends to swallow up both the past and current ones. Let us console ourselves with the thought that for the very reason that what we have said is fragmentary and inadequate, our contemporaries will be all the more inclined to contribute their part toward a complete and finished version.

An Analogous Procedure[107]

Wilhelm von Schütz's[108] *Contributions to Morphology*

JUST AS I am in the process of concluding two volumes, one devoted to natural science in general and the other to morphology in particular, I receive from an esteemed source a publication so useful in its purpose that I cannot refrain from making mention of the pleasure it has given me. It has the title: *Contributions to Morphology*, No. I, 1821, by Wilhelm von Schütz.

The author has apparently comprehended my previous efforts in natural science, has approved my attempt to explain natural phenomena in my own way, and has now decided to transmit what has been revealed to him of the inner and outer world, clarifying and enlivening his account by allusions to events in his own life.

Permit me, once again, to give excerpts presenting me in a favorable light.

"Goethe's contributions to morphology and natural science have exerted undeniable influence upon later publications. The question arises whether indeed the latter would have seen the light of day without them.

"Three characteristics of the Goethean undertaking attracted me strongly and left a lasting impression upon me. I shall first list them, then discuss each one separately.

"Whatever Goethe looked upon in Nature, immediately acquired the character of a living experience for him.

"Treatment and arrangement transform the things he perceives into intermediate entities—wonderful fragments of infinite wholes, both similar and dissimilar—and finished single wholes.

"The originality of content sets the work apart from all previous works of speculative philosophy and contemplative observation of Nature.

"The addition of biographical information puts a historic stamp upon the project. The historical treatment, a welcome tendency of the day, Goethe handles, as he does all things, in the very special manner characteristic of him. In his mind, the inner and outer worlds are more closely united than in other minds, and he appears to be able to characterize as truly historical only what he himself has experienced.

"An individual engaged in a specific subject will experience its his-

---

[107] The publication from which Goethe presents excerpts here was studied by him in August, 1821. The summary was printed in *Natural Science in General; Morphology in Particular*, Vol. I, No. 4 (1822).
[108] Christian Wilhelm von Schütz, 1776–1847.

tory and nature as an inseparable unit by presenting simultaneously the influences of the environment and development of the subject from a previous state. Therefore, since everything historical has its basis in time, all history must become one's own history, to be actually history.

"If Goethe realized this, it quite possibly led him to begin a work which attempted to bring his discoveries in the realm of Nature and in the scientific world into relationship with his personal life.

"A procedure of this kind has a characteristic advantage, for it is often less the subject itself than its formulation that creates a difference of opinion.

"'I wish it to be wholly understood what I have become to Nature and what Nature has become to me,' says Goethe. 'If you wish to understand me even only passably, you must know how Nature found me and I found Nature during our first encounter; then you will have the history and the exposition of my perceptions. It is quite conceivable that this will unite us in the observation of the phenomena to which we are currently devoting our attention.'

"How rewarding indeed it was to understand the historical origin of the science of many things. Yet now that Goethe has led me to this perception, it will not be sufficient to the purpose I have in mind—I must experience things for myself. From this source something wells up, insignificant perhaps in itself, but tinged with a truth capable of replacing many a lack of another sort."

Hereupon, the author speaks as follows of himself and the events of his career that have led him to science:

"Not inclination, but an arbitrary incident permitting no resistance, brought me to the first original activity of mankind, the tilling of the soil. Previously this occupation had been an incomprehensible subject to me. One phase I had failed to understand was the three-field system[109] and everything connected with it, at an age and under conditions where this was inexcusable. However, I was eventually forced to devote myself to this subject to which I was so disinclined. My study was begun with the guidance of the latest textbooks of my acquaintance. Several fortunate insights and coincident combinations of Nature and events rendered my first attempts quite successful, yet my success was due more to good utilization of what the circumstances offered than to the inherent tenability of my method. When I discovered this, I began to rely almost

---

[109] A scientific arrangement in agriculture used especially in the cultivation of grain; in its application a field is planted with grain for two successive years and then allowed to remain fallow for one.

exclusively on observation of Nature, in combination with practical activity of another kind."

For the remainder of this particular incident the reader will undoubtedly gladly refer to the author's own words; and one should by no means be frightened off at the very beginning by a style that is not immediately clear.[110]

"Goethe's individual perceptions of Nature appear by arrangement and treatment as intermediate entities—wonderful fragments of an infinite whole, both similar and dissimilar—and finished single wholes, this was the second statement on our list. It is intended to say that Goethe separates the materials as sharply as though he were treating objects wholly without cohesion. He chooses them with seeming arbitrariness but nevertheless brings them into an unmistakable inner connection, by yielding to scientific, historic, poetical, and philosophical trends of thought.

"Enormous advantage is gained by that separation of single subjects and the exclusion of the extraneous matter which is so prone to insinuate itself into the substance under observation and which nevertheless appears to have its own center in other orbits. The more an author's subject matter deviates from the ordinary mode of thought, the more will he succeed in making his demonstration clear, definite, orderly, and agreeable, for in this way he can penetrate several of the individual things, and the same understanding will surge in upon him that later his reader will experience—to the extent that one individual can draw from the wellspring of another. The two water levels must not be joined in a turbid confluence, and yet neither ought to remain, untapped, within its own sluices."

Here the author reverts to the historical: he mentions Socrates, Aristotle, Plato, Winckelmann[111] and after discussing them, he continues:

"Winckelmann renews the feeling that we have with Plato, that he has not yet completely expressed himself, that indeed he never could or would express himself completely. Goethe is the third intellect of which the same is true, as poet, observer of the universe, as natural scientist. Let us recall here that he is not at all mystical in his poetry, that he never surrounds his subject matter with mystery, but allows the subject matter to approach the reader as a calm, clear, thoroughly understandable and comprehensible natural necessity. At the same time let us remember how close his poetry brings us to the portal beyond which the great world mystery is slumbering, and from which his work took its origin.

---

[110] Regarding the obscurity of Schütz's style, the translator agrees wholeheartedly with Goethe!

[111] Johann Joachim Winckelmann, 1717–1768, archaeologist and art historian.

"This much is grudgingly granted by most individuals. But that the same thing holds true when Goethe takes up the domain of Nature, this only a few will concede. And yet here it is even more remarkable. Let us seek to gain credibility for an assertion as yet not quite clear, by using the following parable.

"Aristotle gives light and Plato soul, but Goethe gives both light and soul when he introduces us to Nature.

"His scientific journal,[112] and especially its manner of arrangement, first revealed this truth to me. The subjects stand out vividly as individual things, by virtue of the method used in the individual essays. Through organization, through combination with biographical matter, through deductions, and through poetic infusion, they are submerged again into the element of a universal soul—the entities are placed against the background of a common whole."

Now our friend proceeds to his last and third point, expressing himself as follows:

"The essential value of Goethe's scientific contributions—and this is the third point to impress us—is closely related to organization and form, for it is precisely its form that prevents the content of individual parts from being torn from the mysterious whole.

"But what is this mysterious whole? What eye has glimpsed it, what authority imparts it, whose thinking has divulged its laws? Isolated things, of course, we do see; isolated results have been embodied in plausible reports; isolated truths have convinced us that they embody undeniable laws; finally and in brief, the isolated is independent of us.

"Let us therefore observe the material. Let us consider its present nature, as it develops before our eyes; let us consider how it came into being. Even though everything assumes a dual form, we shall discover perhaps the unifying principle in Nature, law, and history. Perhaps in the constitution of things we shall find law and history again; perhaps law will show us why history took its particular course and gave rise to a particular effect; history will perhaps give information about the origin of the laws. To prevent ourselves from going astray, to prevent bringing about one result when we intend another and quite different one, let us faithfully, in deep concentration, tarry with the substance in question and not substitute another, even though the latter may often be concomitant. For precisely because each thing customarily represents something beside what it represents in a given occurrence, we must hold it to narrow, definite limits. Nevertheless, for that same reason, we must not

---

[112] *Natural Science in General; Morphology in Particular.*

fail to consider that the same entity which forms our substance may be substance of another sort for other immaterial forces and immaterial requirements. Let us even attempt to make our observation fruitful, by interpreting our results. Such interpretations will be welcome to those who are studying related problems, for naturally the great mystery still looms in the background. Nevertheless, we are convinced that observations devoted to individual things will lead us closer to that mystery.

"These are the most essential and the most important things that appear to be expressed in each of Goethe's latest scientific contributions. The observations of single things and things of our time stand in the foreground, the original phenomena in the background; but these latter are not on that account neglected or too briefly treated, nor do things attainable through reason and intuition suffer curtailment. Even more important is the fact that the sensory fosters the genuinely supersensory.

"Thus the work has the advantage inherent in everything worth knowing, namely, that in limiting hypothesis in the realm of natural science, we will set limits for hypothesis in other fields also and thus provide ourselves with firm and satisfying bases.

"For it is not enough that one or another scientist direct us to this or that basic thought. We must be able to believe in the basis; we must believe that it will meet all our needs.

"A closer description of Goethe's treatment of experiment and hypothesis in his physical and morphological contributions, and its importance for natural history and all other intellectual requirements and tendencies, could easily be added here; but such an extended account will be reserved for a later essay, the present one being merely intended to keep within the limits of general interpretation and not be as yet too extensive."

The foregoing could not be other than extremely pleasing to me. Although it is not desirable that others should imitate us in our actions, it is very gratifying, indeed edifying, when they discover in themselves those human principles by which we act, and then are inclined to chart their own way of life and their own way of imparting ideas.

The Influence of My Publication[113]

THE EARNEST resolve expressed at the end of my treatise on the metamorphosis of plants, not only to pursue this agreeable occupation further

---

[113] The reader is directed in advance to the fragmentary or, as Goethe himself puts it, "aphoristic" character of the following account of the reception of *The*

but also to keep friends of science informed about my current efforts, was blocked during the course of troubled times and in the end was thwarted entirely. At present it would be difficult for me to give an adequate account of the extent to which the central idea of the essay has been influential, and the manner in which it has continued to be discussed down to the very present.

I therefore had to call upon my scientific friends asking them to kindly acquaint me with whatever details they had chanced upon in the course of their studies. Becoming indebted in this way to various persons for information, and finding it possible to collate their notes on individual points, I finally decided to adopt their phrasing too. Consequently this present composition takes on an aphoristic character; this may not necessarily prove harmful, however, as this may be a better way to become acquainted with what has taken place in this field independently, without definite connection.

The first friend to whom I gave some inkling of my thoughts and plans was Dr. Batsch. He entered into the project in his characteristic fashion, not in the least averse to the assignment. And yet the idea seemed to have no influence on the course of his studies, although he was already chiefly occupied in classifying and arranging the plant world into families.

Favorable reviews[114] appear in *Göttingen Review* (February, 1791), *Gotha Scholarly Review* (April, 1791), and *Universal German Library*, Vol. CXVI.

On frequent short visits to Jena, and during one somewhat lengthy sojourn, I often discussed with the excellent scholars of that city scientific subjects which I considered important. Among these scholars, one especially looked with decided favor upon my project: Dr. Johann Christian Stark, Privy Councillor, highly regarded as a practicing physician and in general a brilliant and distinguished man. In accordance with academic tradition he happened to hold the professorship of botany, but only nominally, in his capacity of second-ranking member of the faculty of

---

Metamorphosis of Plants and the further development of the idea presented in it. Although he had the intention of setting off its various parts by differentiated symbols, thus, * and ( ), he failed to carry his plan through consistently. Here, therefore, the sections are spaced apart in the usual manner. The essay—though it hardly deserves the name—appeared first in 1831, as part of the biographical and critical commentary in a new edition of the essay on the metamorphosis of plants. This was the edition with the interleaved French translation prepared under Goethe's direction by Frederic Soret. See footnote no. 149, p. 209; also "Bibliographical Note," p. 255.

[114] See pp. 181–182.

medicine, without ever having taken particular interest in this field. However, the usefulness of my viewpoint could by no means escape a man of his sagacity. Then, after organizing and utilizing in accordance with these views of mine the knowledge he had acquired earlier in the field of botany, he could not resist the temptation to actually discharge the duties of his nominal professorship, and, half in earnest, half in jest, to hold a lecture on botany. By the winter semester of 1791 he had already announced his intention in the catalogue of courses, as follows: *Publice introductionem in physiologiam botanicam ex principiis perill. de Goethe tradet.* For this purpose I turned over to him what I possessed in the way of drawings, copper engravings and dried plants, methodically arranged, so that he was able to enliven his lecture and to carry it out successfully. The extent to which the seeds he sowed at that time may have had a luxuriant growth, somewhere or other, has never been revealed to me. To me, however, such an event seemed to be encouraging proof that discourses of that sort might in future exert a vital influence.

While the concept of metamorphosis was slowly developing in science and literature, I had already had the pleasure of chancing upon a practical man who was fully initiated into these open secrets of Nature.

The aged Dresden court gardener, J. H. Seidel, at my desire and request, produced various plants that had struck my attention in illustrations because of clear evidence in them of metamorphosis. But in asking this favor of him, I did not reveal my purpose.

Placing before me several of the plants I desired, he smiled and said: "I think I see your purpose, and I can produce other examples of that kind, indeed even more remarkable ones." This he did, and both of us were pleased and astonished—I, to learn that he had been training himself to see this great principle at work everywhere in the diversified realm of Nature; and he, to realize that I, as a layman in the field, observing zealously and earnestly, had developed the same technique.

In a heart-to-heart talk he revealed the rest of the story, explaining that this concept had enabled him to interpret much that was difficult to understand, and that he had also been able to apply it practically.

Just how this essay[115] has influenced the course of science in Germany up to the present, is a highly complex question which probably will not be answered adequately until the conflict of opinions has calmed down and the combatants regain their senses. For it really seems to me as if the idea of metamorphosis has laid hold of many who are not aware of it, while others who preach the new thesis know not whereof they speak.

---

[115] Goethe's essay on the metamorphosis of plants.

Nothing seems to be more difficult than the progress of an idea new to science to the stage where it is included in systematic instruction, in that way giving proof of its vitality. Let us now present those successive stages in greater detail.

Dr. Friedrich Siegmund Voigt[116] based his botanical lectures of 1803 on these observations and made mention of them again in the first edition of his botanical dictionary of the same date. In his preface to *System of Botany*, 1808, he gave a detailed free rendering of my work in a special section.

At the same time, the idea of metamorphosis was given definite recognition and expertly applied in the further development and transformation of science in Kieser's *Aphorisms of Plant Philosophy*, 1808. On page 61 it is explicitly stated, after a discussion of Linné's theory of anticipation: "From this, Goethe created with originality and brilliance a general theory of metamorphosis, and this theory has remained for years the most comprehensive treatment of the special subject of plant physiology." We must not judge this work, so closely linked with Schelling's philosophy, on the basis of its present-day aspect. In its time it created a sensation; and rightly so, for it is rich in original ideas drawn from the depths of Nature.

In 1811 Friedrich Siegmund Voigt published a brief work, *Analysis of Fruit and Seed-Corn,* in which he already betrayed his annoyance that up to that time no botanist had yet had the desire to agree with this theory. His own words, page 145, are: "I shall therefore refer at once to Goethe's indisputable theory of metamorphosis, which has been ignored by some scientists from sheer spite [my essay cited in a footnote]. In the theory it is shown through examples of all kinds how the plant starts the highest organs toward their life goals, by means of expansion and gradual contraction. These highest organs are, as has been said, merely the original ones which have become more complex and which have changed color through repetition of the selfsame act of transformation.... The discussion of metamorphosis in the flower system is limited principally to the manner in which the leaves are transformed. However, from the very first type of development onward, the famous creator of that theory has drawn attention to still another formation, that of the nodes."

With the year 1812 another instance of recognition comes to our notice in a book which, in fact, owes its basis, its very existence to the theory of metamorphosis: G. Fr. Jäger's book on the malformation of

---

[116] See pp. 121, 122, 189, 201, 206.

plants. We find this statement on page 6: "In both types of propagation the further development of the new individual takes almost the same course, which generally consists in a constantly progressing formation of new organs up to the flower. This flower, even though complete in itself, reveals in the structure of its own organs its relationship to the other organs, so that they all appear to have originated one from the other through metamorphosis. On this subject we are indebted to Herr von Goethe [my publication cited] for a more detailed presentation, in which he has also considered individual instances of plant retrogression."

It can undoubtedly still be remembered that Schelver[117] meanwhile had based his *Critique of the Theory of the Sexes of Plants* (1812) entirely on metamorphosis and that the dispute thus engendered had taken the upper hand and had degenerated into abuse. Had the worthy author not become embittered, first through unseemly treatment of himself and then through rash overrating of his pupil,[118] from which a retreat was soon made; had an agreement been reached, instead, on the concept of plant individuality, which was all-important since Schelver based his argument on the impossibility of hermaphroditism in the individual: I am convinced that in this way too the theory of the sexuality of plants would have been rescued, purged, and strengthened. Wind and insects would have been abandoned, amply compensated for by metamorphosis. But regardless of the manner in which the dispute was carried on, metamorphosis at least had to be mentioned often; nothing more was needed to win adherents for it even among Schelver's opponents. Young Authenrieth[119] is one of these adherents.

Powerful influences were undoubtedly the new German philosophy on the one hand, and the gradual ingress of the natural plant system among us until a point of entry was created for the theory of metamorphosis. That theory was bound up in turn with the study of plant geography, which had become a favorite field since Humboldt's return and which is so inseparable from the natural plant system that even the most obstinate adherent of Linné, even Wahlenberg,[120] had to make the best of it by at least leaning on the old *Ordines naturales* of Linné.

Lasting influence was exerted by Kieser's[121] *Mémoire sur l'organisa-*

---

[117] For previous references to Schelver, see pp. 105, 106, 119, 189, 241.
[118] Henschel; see p. 106.
[119] Hermann Friedrich Authenrieth, professor in Tübingen.
[120] Göran Wahlenberg, 1780–1851, Swedish botanist and geologist.
[121] Dietrich Georg Kieser, 1779–1862, professor of medicine, Jena. See p. 241.

tion des plantes, 1814, and a German abridgement of this lengthy work in 1815. Of these publications, also, it may be said that metamorphosis is not merely grafted to a finished trunk but forms its very substance and soul. And since these two publications are based more directly on observation, the peculiarity of the school of thought which the author embraces obtrudes less offensively for those who hold opposing opinions. In France, to be sure, no attention was paid to Kieser until recently, after the dictatorship of Brisseau-Mirbel,[122] his outright opponent, had been broken up by Dutrochet[123] and others. In Germany, however, he soon acquired such authority that Treviranus,[124] and a few others who still remained impartial, were able to make but slow headway, even against Kieser's obvious mistakes. Even in Nees von Esenbeck's *History of Botany,* 1820, the anatomical investigations of Moldenhawer,[125] Treviranus, and others appear to be slighted somewhat in favor of Kieser's.

Then Nees von Esenbeck[126] attempted to enlarge the territory of the metamorphosis theory in another direction. Even in the simplest, leafless plants *(Fresh-Water Algae,* 1814; *System of Fungi,* 1815) he sought to demonstrate metamorphosis and to arrange it according to stages. His later *Botanical Handbook* rests on the same fundamental views, which, if they do not coincide exactly with those originally expressed by Goethe, are at least quite similar to them, and by von Esenbeck's own statement are gratefully derived from this source.

In addition, most of the credit for disseminating this more natural and vivid conception of plant formation is due to this excellent man. He exerted extraordinary influence through his careful editing of the proceedings of the Leopold-Caroline Academy, through vigorous participation in the *Regensburg Botanical Gazette* and other journals, through reprinting Brown's writings in the original and in translation, through correspondence and personal instruction.

Friedrich Siegmund Voigt comes forward boldly with his *Outlines of Natural History* (1817 and after), and beginning on page 433 gives

---

[122] Charles François Mirbel, called Brisseau-Mirbel, 1776–1854, the first strict representative in France of plant anatomy based exclusively on microscopy.

[123] René Joaquim Henri Dutrochet, 1776–1847, French botanist who sought to explain the total physiological processes of plants, beyond purely anatomical data based on microscopy.

[124] Christian Ludolf Treviranus, 1779–1864, professor of botany, not to be confused with the natural philosopher R. Treviranus.

[125] Johann Jacob Paul Moldenhawer, 1766–1827, professor of botany, Kiel.

[126] See also pp. 104, 241.

a free rendering several pages in length from that publications,[127] graphically illustrated by a copper engraving of *Helleborus foetidus.*

Kurt Sprengel in his *History of Botany,* 1818, II, 302, expresses himself as follows: "Von Goethe demonstrates most clearly and interestingly the development of plant parts from one another. [Passage cited.] Provision is made for development through compression of the forms; this principle of vegetation Goethe amplifies in convincing and instructive manner. . . . That the nectaries are usually transitions from petals to stamens, that even pistil and stigma through retrogression become similar to petals and merely develop from the latter by contraction, becomes obvious when the stamens come to resemble the petals in cases where the latter miscarry (in several varieties of *Thalictrum*). This brilliant scientist evidently felt that malformation and doubling of flowers were very useful to his theory and, therefore, returns to these subjects.

"Goethe's metamorphosis had too deep a meaning, was so satisfying because of its simplicity, and was so teeming with useful conclusions, that it is little wonder it gave rise to further discussion; some, however, insisted on ignoring it. One of the first to adopt Goethe's ideas in a textbook was Friedrich Sigmund Voigt, professor in Jena *(System of Botany,* 1808). Very interesting ideas on the relationship of stamens and petals, as well as on the predominant numerical proportions, were presented by Johann Ludwig Georg Meinecke[128] *(Proceedings of the Society of Natural Scientists in Halle,* No. 1, 1809). Oken[129] also treated in detail the principle of metamorphosis in his *Natural Philosophy.*"

In the same year [1818] an essay, of which Nees von Esenbeck was apparently the author, appeared in the periodical *Isis,* p. 991. It is entitled "On the Metamorphosis of Botany," and the historical introduction of the subject begins thus: "Theophrastus was the creator of modern botany; Goethe has become its gentle, friendly father. Now, the daughter, beginning to experience human emotions and love, and to develop beauty of form, will look up to him more and more fondly as she emerges from childhood and recognizes more fully the value of her own lovely existence and of the paternal care he has given her.

"J. W. von Goethe's *Essay on the Metamorphosis of Plants,* Gotha, Ettinger, 1790, 86 p., is urged upon us even more strongly by the first

---

[127] Goethe's essay on the metamorphosis of plants.
[128] Johann Ludwig Georg Meinecke, 1781–1823, professor of technology in the university of Halle.
[129] See pp. 128, 241.

number of a new series of scientific treatises, with the general title: *Natural Science in General."*

Dr. H. F. Authenrieth[130] *(Disquisitio quaestionis academicae de discrimine sexuali jam in seminibus plantarum dioeciarum apparente, praemio regis ornata,* Tubingae, 1824) is acquainted with the theory of metamorphosis and touches on it, page 29, with the words: "The way in which the sexual organs of the two sexes are formed in the hemp plant coincides completely with what Goethe has already expounded; therefore I feel impelled to mention that I have seen instances in which the anthers, and also the seeds with their pistils, have originated from sepals."

Although we should possibly have mentioned Dr. Ernest Meyer earlier, it is quite in order, from the point of chronology, to mention him at this particular point. He is at present full professor at the University of Königsberg and director of the botanical garden there.

I was never privileged to have personal relations with him, but his sympathy and interest encouraged me from the very start. Without specific and detailed mention of metamorphosis in his writings, this friend has advanced my thesis both through pure theory and zealous preaching. Proof of this is afforded by the important work cited below, by one of his pupils, which it will give us great pleasure to discuss.

Röper's[131] *Enumeratio Euphorbiarum* is one of those rare works which, although written entirely in consonance with the theory of metamorphosis, rarely mention it, and therefore find readier acceptance among those who hold opposing opinions. Its material also was preeminently suitable for such treatment. Richard,[132] the actual author of Michaux's[133] *Flora boreali-americana,* has already shown that what Linné had regarded as single flowers of *Euphorbia,* might also be considered as an inflorescence, or *flos compositus;* the supposed pistil might be considered as the central female flower, the allegedly jointed stamens as a verticil of single male flowers with stems; the corolla as an involucrum; and so forth. Later, by comparison with the structure and development of related species, Robert Brown, and likewise Röper, attempted to confirm that view, particularly by using numerous instances of highly striking malformations.

---

[130] See p. 199.
[131] Johann Röper, 1801–1885, professor of botany in Basle and Rostock.
[132] Louis Claude Marie Richard, 1754–1821, professor of botany in Paris.
[133] André Michaux, 1746–1802, botanist.

In 1823 we received an excellent work: *Lud. H. Friedlaenderi*[134] *de institutione ad medicinam libri duo, tironum atque scholarum causa editi*. In the course of brilliant advice on acquiring a thorough knowledge of medicine, he devotes several paragraphs to botany, stating on page 102: "Thus there is nothing free or arbitrary about plant growth; instead, we find that the life activity of the plant is definitely planned and is directed solely toward growth. This growth is effected partly through expansion and partly through contraction, in such a manner that the root takes a downward direction from the developed embryo, and the stem an upward one, the latter finally acquiring the ability to produce, from a succession of leaves, the calyx, corolla, stamen, and fruit organs, indeed even the fruit itself."

It is now the fashion in all botany textbooks, which will soon be legion in number, to assign a small chapter to metamorphosis. But the spirit which should animate and penetrate the whole, cannot be thus confined. Writings of that sort will be omitted here, for they are only consulted by beginners in the hope of finding some technical expression they are in need of.

H. F. Link,[135] *Elementa philosophiae botanicae*, Berolini, 1824. The author says, p. 244:

"The metamorphosis of plants has been admirably expounded by Goethe. He shows the plant alternating between expansion and contraction; the flowers can be regarded as an instance of contraction; yet although this is true of the calyx, the corolla on the other hand is in a state of expansion. The stamens, anthers, and the pollen in turn represent the greatest contraction; the pericarp on the other hand is again expanded, and so on, up to the highest form of contraction in the embryo. This oscillation of Nature occurs not only in mechanical movements, such as those of the pendulum, waves, etc., but also in living bodies and in the rhythms of life."

This apparent praise of our efforts was bound to seem suspicious to us, for where form and transformation ought to have been discussed, only an ultimate, idealized, sublimated abstraction is cited, and highly organic life is adjoined to completely formless and incorporeal natural phenomena of the most general nature.

Our adverse feeling was increased to the point of depression when

---

[134] Ludwig Hermann Friedländer, 1790–1851, medical man.
[135] Heinrich Friedrich Link, 1767–1851, professor of botany, director of the botanical garden in Berlin.

we discovered on most careful examination that these words had been squeezed into this work, complete strangers and intruders, and definitely robbed of all meaning. For at the very outset and throughout his demonstration (see index) the author not only uses the term metamorphosis in an entirely different sense than we and others use it, but also in cases where it does not even suit his own purpose. For how is one to interpret the closing words, page 152: *Hoc modo nulla fit metamorphosis!*[136] In addition, he adds each time a so-called anamorphosis, in that way obscuring the real meaning.

Most deplorable, however, is his desire to trace back the chief and final formation in flower and fruit to Linné's untenable theory of anticipation, for which he would need not *one* but a dozen prolepses; furthermore, because of the proleptical use of latent annual buds, he necessarily confines himself to enduring trees, adding quite naively: *Ut prolepsis oriatur ligno robusto opus est*[137] (p. 246).

But what happens in the case of an annual plant that has nothing to anticipate?

It is our opinion that the transitory plant, soon to fall into decay, is enabled by rapidly intensifying metamorphosis to produce in advance hundreds and thousands of possible plants, which, to be sure, will likewise be ephemeral but potentially and immeasurably fruitful in addition. Not a *prolepsis* of the future plant but a *prodosis* of generous Nature, is the term we should apply, and in this correct expression find both pleasure and instruction.

Enough! Or rather, too much! We should not quarrel with the erroneous. Merely to point it out should suffice.

In this series we should give honorable mention to a name of importance, that of Robert Brown.[138] It is characteristic of this great man rarely to mention the fundamental truths of his science, and yet each work of his shows how intimately acquainted he is with those truths. Hence the complaints about his obscurity of style. On the subject of metamorphosis also he has never fully declared himself. Only once, in a casual note in his essay on *Rafflesia*, does he state explicitly that he considers all flower parts as modified leaves, attempting then to explain the normal formation of anthers from this point of view.

Those casually uttered words of the botanist recognized as the greatest of our time have fallen on fertile soil and have had profound influence

---

[136] "No metamorphosis results in this manner!"
[137] "Solid wood is required for the occurrence of prolepsis."
[138] Robert Brown, 1773–1858, famous English botanist. See also p. 200.

especially in France. Aubert du Petit-Thouars in particular, extolled by Brown as one of the defenders of the metamorphosis theory, appears to owe to Brown's favorable opinion, expressed here and elsewhere, the esteem which he is beginning to enjoy in France, and which he was unable to win directly from his prejudiced countrymen through his achievements.

A. P. de Candolle,[139] *Organographie végétale*, 2 tomes, 1827, Paris. In speaking of the work of this outstanding man we prefer to cite several[140] passages from other authors. Our translator, de Gingins-Lassaraz,[141] says the following in his historical preface to our essay on metamorphosis:

"Meanwhile a famous botanist, unacquainted with Goethe's work, attacked the subject in his own way. In 1813, guided by a brilliant talent, which I shall not venture to evaluate fully, and supported by profound knowledge of the plant realm and an immense amount of experimentation and observation, he expounded in his work *Elementartheorie* the principles of the symmetry of organs and the history of their metamorphoses—which he called "degenerations." This theory, based on these solid premises, had no need to fear the fate of Goethe's work; practically and theoretically it made numerous and swift advances in the treatment of plants and was completed by *Organographie végétale*, which summarizes all our knowledge on the subject."

P. J. F. Turpin.[142] From this outstanding man, celebrated both as a brilliant botanist and painstaking draughtsman of full grown plants as well as their microscopic beginnings, we have appropriated a motto, which we found below Plate I, Vol. XIX, of *Memoirs of the Museum of Natural History* in 1830 and which we should like to repeat here because of its importance: "Watching things develop is the best means to understanding them." Elsewhere he states: "The organization of a living creature in general, and that of its organs in particular, can be understood only by following its development step by step from the moment of its first emergence as a form to the moment of its death." And this also remains the chief article of faith for Germans who are serious in their endeavors to observe Nature faithfully.

The graphic artist, commissioned to reproduce faithfully the smallest variations among the subjects assigned to him, will very soon discover,

---

[139] See pp. 10, 165, 209.
[140] Only one is cited.
[141] Frédéric de Gingins-Lassaraz in 1829 had published a translation of Goethe's essay on metamorphosis. See p. 208.
[142] Pierre Jean François Turpin, 1775–1840, author of a great iconography of plants.

as he transmits them to paper with a practiced hand, that the organs of any particular plant are not clearly differentiated. He will soon become aware of gradations from one organ to another and of their climatic development, and it will then be easy for him to draw skillfully the constant succession of organisms which are always the same and yet always different.

Among the words for which we must envy the French language is the expression *s'acheminer*. Though its original meaning was merely "to set out on one's way," the ingenious French must have felt that the traveller's every stride had a different meaning, a different import from the preceding one, in that the goal to be reached is perfectly understood and contained in each individual step when once the correct route has been chosen. For this reason the term *s'acheminer* has assumed an ethical connotation. In using it one thinks of "advancing" and "progressing," but in a higher sense. Indeed, all strategy rests fundamentally on the most opportune and vigorous *acheminement*.

Turpin frequently had opportunities of applying to plants the best of what has just been said, not only through scientific observation but also through artistic representation in drawings. He would therefore be rendering the utmost service in this field if he were to apply his skill in a graphic representation of plant metamorphosis.

To be sure, the plates in the *Organographie* by the sharp-sighted de Candolle have already given us strikingly instructive examples of this. However, for the special purposes we have mentioned we should like to have a fuller and a strictly accurate assortment, arranged methodically in natural sequence, and above all, given characteristic emphasis through use of colors. This should not be too difficult a task considering the decidedly botanical cast of mind of this outstanding artist and the helpful preliminary work he has already done.

If we had the good fortune of living near this consummate artist, we would daily and urgently beseech, implore, and challenge him to undertake this task. Such a work would require very little explanatory text and would supplement botanic terminology and its abundant vocabulary; yet such a work would have an independent existence also, for this original language of the natural elements, extensively elaborated and applied, must inevitably seem completely understandable to us.

In 1827 the second edition of Friedrich Siegmund Voigt's[143] *Botanical Textbook* made its appearance. On pages 31 and following, an expla-

---

[143] Entitled *System of Botany* in its first edition; see p. 201. For other references to Voigt, see pp. 121, 122, 189, 198.

nation of metamorphosis is reprinted as it originally appeared in the first edition, except that in this new edition the discussion is more closely linked with basic principles of botany and is supported by numerous examples from little-known works and from personal observation.

Botany for Ladies, Containing a Discussion of Metamorphosis of Plant Life, by Ludwig Reichenbach,[144] Leipzig, 1828. The author, after presenting the views and methods of Linné and Jussieu, turns to my work, stating as follows:

"Goethe probes deeply into the life of Nature, and his expert interpretation of what he has observed, his felicitous interpretation of particulars in the context of the whole, and in general the originality and sweep of his views on Nature, lead us to recognize in his efforts a third direction which natural science is in a position to take. And especially because he devotes so much attention to the all-important observation of the plant world and to investigation of its development and growth, we must justly say of him that although he was already probing the secrets of the dryads in his youth he had to wait until old age before the world understood him! His ingenious essay on the metamorphosis of plants (Gotha, 1790), characterized by a power of observation as brilliant as the interpretation is vivid and felicitous, did not attain the high fame it merited until very late. This theory of metamorphosis, of the development of the individual plant, when extended to the entire plant realm, provides laws for an ideal classification and for a vivid description of that connecting link for which we must make eternal search without ever being able to reach it. The writings of this great master merely furnish a pregnant clue; the solution is open to each individual in proportion to his insight, zeal, and vigor."

As evidence of our complete approval of the effort of this brilliant man, we append these few words: An idea, the moment it finds expression, becomes a splendid public treasure. Whoever succeeds in capturing it, wins a new possession without robbing anyone else. He makes consistent use of it in his characteristic fashion, moreover, even without having it constantly before his mind. And this precisely is testimony to the enduring worth and strength inherent in the property acquired.

The author dedicates his work to women, artists, and thoughtful lovers of Nature; he hopes that his work will encourage the study of this lofty principle in Nature and the application of it in practice. May he find a most happy reward through its success!

---

[144] Heinrich Gottlieb Ludwig Reichenbach, 1793–1879, professor of natural history and director of the museums of Dresden.

Botanical Review, Vol. II, No. 3, Nürnberg, 1829 (p. 427); Royal Institute of Great Britain in London, 1829. On January 30, among others, Mr. Gilbert T. Burnett[145] read a long essay on plant metamorphosis. Only an excerpt is printed here, and it would be desirable to have the whole essay before us. It apparently does not fully coincide with my own views, but gives the subject serious and intelligent treatment.

A French translation of the *Essay on the Metamorphosis of Plants* will undoubtedly exert a beneficial effect. Its central idea has been stirring on the other side of the Rhine also; Aubert du Petit-Thouars[146] and Turpin (in his supplement to Poiret's[147] *Leçon de flore*) give the clearest proof of that. Yet, in my opinion, both have already strayed far beyond justified limits and have found little approval among their countrymen. Let us hope that this simpler and more natural treatment will reconcile some and, conversely, call others back to the right track.

Essai sur la Métamorphose des Plantes, par J. W. de Goethe; traduit de l'allemand sur l'Edition originale de Gotha (1790), par M. Frédéric de Gingins-Lassaraz, Geneve, 1829. In a historical preface the translator makes the following statement: "There are two very different methods of studying plants. The most common is to compare with one another all the individual plants making up the entire world of vegetation. The other method compares the various organs comprising the individual plant and searches there for the characteristic symbol of plant life. The first of these two methods of study leads us to a knowledge of the plants that exist throughout the world, and of their environment, mode of life, and uses. The second method acquaints us with the organs of the plant, with their physiological functions and the roles which they must play in the plant economy. It studies the course of development, the metamorphoses to which the individual parts must adjust themselves; it allows us to see the plant as an organism which is born, grows, reproduces, and dies. In brief: the one method is the history of plants; the other, the history of *the plant.*

"This latter method of considering plants has been called the philosophical because it is more closely associated with natural philosophy. Actually, these two methods of studying living organisms are quite inseparable. It would be absolutely impossible to understand the natural

---

[145] Gilbert Thomas Burnett, 1800–1835, professor of botany, London.

[146] Louis Marie Aubert du Petit-Thouars, 1758–1831, French botanist and traveler. See p. 205.

[147] Jean Louis Marie Poiret, 1755–1834, French botanist and explorer of Africa.

relationships of the plants being compared if there were no means of evaluating the various disguises which the organs assume before our very eyes; and on the other hand, the true nature of the organs themselves can be disclosed to us only if we compare analogous parts of a great number of varied plant genera.

"These considerations, we hope, will dispose readers favorably toward our translation, and gain a wider public for Goethe's brilliant essay on plant metamorphosis, inasmuch as the passage of time and careful observation of the subject have more or less proved the truth of his theory.

"It was the privilege of this author, whose literary productions are well known for their free and natural style, to turn his brilliant mind also to the plant realm, where, free of dogma and prejudice, he shows us the plant in all its simplicity, silently and mysteriously exercising its capacity to grow, to flower, and to reproduce.

"The author, curbing the natural ardor of his imagination, relying instead on a small number of easily accessible but well-chosen examples, undertakes to guide his readers step by step on a path as simple as it is clear, to a conviction of the truths with which he himself is imbued. Moreover, his theory is elementary in the highest sense of the term and well-suited to teach and convince those who have not actually studied plants. And in this respect it might serve as an example to those individuals desiring to give wider circulation to knowledge of the organisms round about us and, as the expression goes, to popularize it."

The afore-mentioned translation,[148] since the original[149] is again included here, will now be judged by scientific experts of both languages, and it will be found that in the interest of clarity such expressions were used for the most part[150] that appertain to the present state of science.

The present translator[151] takes the liberty of saying the following about his efforts: since he had the good fortune several years ago to attend lectures by de Candolle and has not in the meantime neglected the study of natural history, he is not entirely unacquainted with the principles[152] of plant physiology and the terminology pertaining to it.

---

[148] Thus, that of Gingins-Lassaraz.
[149] Namely, Goethe's *Metamorphosis* of 1790 itself, which "here" was being newly published in German, with an interleaved French translation prepared by Frederic Soret, 1795–1865, under the direction of Goethe himself.
[150] The reference now is again to Gingins Lassaraz.
[151] Soret.
[152] With those of 1831, in contrast to those prevailing in 1790.

However, he was able to regard these accomplishments as more dangerous than advantageous inasmuch as the first exposition of the theory of metamorphosis is much earlier[153] than the creation of the present physiological nomenclature. Thus, he often carefully avoided those expressions which are in general use at present, in order to render more faithfully the meaning of the original, by an exact translation done under the eyes of the author.

Reichenbach's work is announced in the *Bulletin des sciences naturelles, sous la direction de M. le Baron de Férussac,*[154] No. 5 (p. 268), Mai 1830. *Botanique pour les dames, les artistes et les amateurs des plantes, contenant une exposition du regne végétal dans ses métaphores* (sic!) *et une instruction pour étudier la science et pour former des herbiers.*

To this translation of the title nothing further is added, not even the slightest hint as to the possible content. In an announcement, directly following, of a German work on natural science, the reviewers state that they list the two works merely to avoid omitting anything that is being printed on any scientific subject whatsoever.

Now it seems to us that the editors should have felt obliged to append a few remarks to their announcement of the above-mentioned book, in view of the long-standing influence of the morphological theory in Germany, a theory which has long since been introduced into France by an acknowledged master in the field,[155] and which has recently even been revived through a translation of our first essay on the subject.

However, as far as the odd misprint is concerned, which distorts the above-named title through printing "metaphors" instead of "metamorphosis," we consider our contemporary world too erudite to suspect in it a sarcastic allusion to the German manner of handling scientific subjects! The theory of metamorphosis cannot be unfamiliar to the editors, and they will regret that they did not read proof more carefully, or, more likely, that they entrusted the editing and revision of this chapter to individuals who were complete strangers to the scientific profession.

J. P. Vaucher,[156] *Histoire physiologique des plantes d'Europe, ou exposition des phénomenes qu'elles présentent dans les divers périodes de leur développement,* Geneve, 1830.

---

[153] Namely, in 1790.

[154] André Etienne Just Pascal Joseph François d'Audebard, baron de Férussac, the younger, 1786–1836; zoologist, whose specialty was conchology.

[155] Undoubtedly de Candolle.

[156] Jean Pierre Etienne Vaucher, 1763–1841, botanist and clergyman in Geneva.

Although we have found this important work useful from its very publication, we would not actually need to mention it here. The author, an intelligent botanist, explains natural phenomena from the standpoint of teleological views which are not and could never be ours, although we have no quarrel with the individual who employs them.

But inasmuch as the author announces at the close of his introduction that he is disinclined toward the methods which Herr de Candolle uses to develop the subject of botanical organization in his didactic writings, and inasmuch as he thus simultaneously rejects our own view, which almost coincides with de Candolle's, we avail ourselves of the opportunity of discussing these admittedly subtle relationships.

To be sure, one must be grateful that such an outstanding man as de Candolle recognizes the identity of all plant parts and presents such varied examples of their agility and mobility when progressing or retrogressing in form, thereby presenting themselves to the eye in infinitely diversified form; however, we cannot approve the course he takes in guiding the amateur botanist to the fundamental and all-important idea. In our opinion he does ill to proceed from *symmetry,* and to even designate the process by this name.

The good man assumes a certain intentional regularity on the part of Nature, and everything that fails to coincide with it he calls abnormalities and deviations, which—through abortion, extraordinary development, atrophy, or amalgamation—mask and conceal that basic principle.

It was precisely this manner of speech that intimidated Herr Vaucher, and we cannot exactly blame him. For the actual goal of Nature would consequently seldom be attained in the plant world; we would be shunted from one exception to the other and nowhere would we find firm footing.

Metamorphosis is a higher concept—it governs the regular and the irregular; it explains the formation of the single rose as well as the double one, the production of the common tulip as well as the rarest orchid.

In this way all success and all failure of Nature's products are clarified for the skilled scientist. The eternal flexibility of life becomes clear to him; the possibility exists that plants may develop in most unfavorable as well as favorable conditions and that species and varieties can extend to all zones.

When a plant, through inner laws or through pressure of exterior causes, changes the form or the relation of its parts, it should be looked upon as being in absolute accordance with law, and none of these deviations should be regarded as misgrowth or retrogression.

Whether an organ lengthens or shortens, expands or contracts, fuses or splits, develops prematurely or too late, displays or conceals itself, all takes place in accordance with the simple law of metamorphosis. Its operation is revealed in the symmetrical and the bizarre, the purposeful and the ineffectual, the conceivable and the inconceivable.

A demonstration of this kind would perhaps be more acceptable to Herr Vaucher, were it possible to discuss the subject with him step by step and to submit examples in proof, since the teleological view is aided rather than refuted thereby.

More and more evidence is available to the scientist to prove that things meager and homogeneous are capable of producing the greatest diversity when set in motion by the eternal primordial force.

The careful observer can glimpse the seemingly impossible even with the unaided eye, a fact which forces one to prostrate oneself in adoration before the mysterious origin of all things, call it considered purpose or inevitable consequence, as one will.

Were I addressing Germans alone, I should continue, conversing pleasantly and openly in a language understandable to them, as to kindred spirits. However, since I expect the French translation to appear on the opposite page,[157] I prefer to stop here lest I be suspected of mystic musings by a nation which demands everywhere, and from every one, complete clarity in thought and expression.

---

[157] Goethe is referring to the interleaved French translation. See footnote 113, p. 195 f.

On General Theory

EPIRRHEMA

In contemplating Nature's being,
Know the One as many, seeing
In and outer coinciding,
Nothing in from out dividing.
Open secret, revelation!
Grasp it without hesitation.

Free of seeming truth's confusion,
Revel in the serious game!
Separateness is the illusion—
One and many are the same.

—Goethe

(English rendering by Aldyth Morris)

Propitious Encounter[1]

THE HAPPIEST MOMENTS of my life were experienced during my study of the metamorphosis of plants, as the sequence of their growth gradually became clear to me. This method of regarding the plant world inspired me during my stay at Naples and Sicily; it became more and more precious to me; everywhere I gave myself practice in its application. And these pleasant pursuits were to achieve priceless value by providing the occasion[2] for one of the noblest relationships granted me in my later years. For it was my interest in these phenomena that led to a more intimate association with Schiller, bringing to an end the unfriendly relations that had kept us apart so long.

On my return from Italy—where I had sought to foster in myself greater precision and purity in all branches of art, unaware of what had meanwhile taken place in Germany—I found certain poetic works, new and old, enjoying great influence and popularity at home, unfortunately of such a sort as to inspire in me the utmost repugnance. I shall mention only Heinse's *Ardinghello*[3] and Schiller's *The Robbers*. The first-named author was abhorrent to me because he had used creative art to refine and adorn sensuality and illogical methods of thinking; the second, because a vigorous though immature talent had poured out over the fatherland, in ecstatic torrents, precisely those ethical and dramatic paradoxes of which I had been endeavoring to purify myself.

I did not blame the two men themselves for what they had undertaken and achieved, for no man can gainsay the promptings of his nature. At first he accepts these dictates unconsciously and naively, then more and more consciously at each stage of his development—this being indeed the explanation why so many excellent and foolish things alike are spread over the world and why confusion develops from confusion.

However, the furore these writers had created in my country, the acclaim they were receiving from undisciplined students and cultivated court ladies alike, appalled me. It seemed to me that my striving had

---

[1] Written in May, 1817, this account appeared in *Natural Science in General; Morphology in Particular*, Vol. I, No. 1 (1817).
[2] In 1794.
[3] Wilhelm Heinse, 1749–1803, minor German writer, author of the popular Renaissance novel mentioned here.

all been wasted. The pursuits in which I had educated myself, the manner and style in which I had done it, seemed shunted aside and invalidated.

The thing that pained me most was that my friends, Heinrich Meyer[4] and Moritz,[5] and likewise the artists that were working on the same principles, seemed to be endangered: I felt much concern. Had it been possible, I would gladly have given up my study of the plastic arts and the exercise of my literary talent completely. How could I hope to be heard in competition with those rhapsodical, extravagant works? Let the reader picture my position. Long endeavoring to nourish and support the purest conceptions, I now found myself wedged in between an Ardinghello and a Franz Moor.[6]

Moritz had also returned from Rome and was spending some time with me. We lent each other moral support and passionately fortified our mutual convictions. Schiller, though he lived near me during his stay in Weimar, I avoided. The appearance of *Don Carlos* was not likely to draw us closer together. I resisted the attempts of mutual friends to bring us together, and thus we continued to live side by side, yet strangers.

His essay, "On Grace and Dignity in Literature," was as little calculated to conciliate me, for in it he rapturously embraces the Kantian philosophy, one which elevates the subjective to great heights while appearing to circumscribe it. The essay plainly revealed the extraordinary gift Nature had bestowed upon him. Yet with immoderate feelings of emancipation and self-determination he played the ingrate to the Good Mother, though she certainly had not played the stepmother to him.[7] Instead of independently and actively observing her manner of creating, as she advances according to law from the lowest to the highest forms, he approached her from the standpoint of a number of empirical human traits. Certain harsh passages I could even interpret with reference to myself, but though they showed my convictions in a false light, I felt it would be even worse if they had not been said with me in mind, for

---

[4] Johann Heinrich Meyer, 1759–1832, Swiss painter and art historian, one of Goethe's closest friends. They had met in Rome, and later Goethe was instrumental in his going to Weimar, where he taught art and later became director of the Art Institute.

[5] Karl Philip Moritz, 1757–1793, philologist and archaeologist.

[6] Central figure in Schiller's *The Robbers*.

[7] The passage Goethe is alluding to here is undoubtedly the following: "Even animal forms speak, inasmuch as their exterior form reveals their interior. Here, however, it is only Nature that is speaking, never Freedom." Only Nature! As one critic points out, it was as though Schiller had said: "Thou mayest honor other gods, it is *only God* who has forbidden it." (Richard M. Meyer, *Goethe*. [Berlin: Ernst Hofmann & Co., 1905], I, 377.)

then the enormous chasm between our modes of thinking must be even more unbridgeable.

Reconciliation seemed out of the question, and even the mild persuasion of a Dalberg,[8] who had given Schiller the high evaluation he deserved, had no effect. Of course, the reasons I advanced against a reconciliation were hard to refute. No one would deny that in the case of intellectual antipodes more than just the distance of the earth's diameter constitutes the separation, and that, as poles, they cannot actually ever coincide. However, a relation may nevertheless exist between them, as proved by the following incident.

Schiller had moved to Jena, where, as before, I saw nothing of him. It was about this time that Batsch,[9] with unbelievable enterprise, had founded a scientific society, with fine collections and impressive apparatus. I usually attended its periodic meetings and one time found Schiller there. By chance we left the hall together, and began a conversation. He appeared to be interested in the lectures, but remarked with great insight, and to my pleasure, that such mangled methods of regarding Nature could only repel a lay person who might otherwise be willing to venture into the subject.

I answered that perhaps even to experts such a method would be uncongenial and that there might be another way of considering Nature, not piecemeal and isolated but actively at work, as she proceeds from the whole to the parts. Schiller expressed the desire to have the point clarified through discussion, though not concealing his doubts and refusing to grant that my views owed their origin to experience.

We had reached his house; the conversation lured me in. I gave a spirited explanation of my theory of the metamorphosis of plants with graphic pen sketches of a symbolic plant. He listened and looked with great interest, with unerring comprehension, but when I had ended, he shook his head, saying, "That is not an empiric experience, it is an idea." I was taken aback and somewhat irritated, for the disparity in our viewpoints was here sharply delineated. The statement from his essay, "On Grace and Dignity in Literature," occurred to me again; the old antipathy was astir. Controlling myself, I replied, "How splendid that I have ideas without knowing it, and can see them before my very eyes."

Schiller, who had far greater tact and urbanity than I, who, furthermore, in the hope of procuring my help with his magazine *The Hours*,[10]

---

[8] See p. 175.
[9] See pp. 76, 96, 155, 183, 189, 196 for other references to Batsch.
[10] *Die Horen,* publication named for the Greek goddesses of time, published by Schiller, 1794–1797.

was desirous of attracting rather than repelling me, replied in the manner of a trained Kantian. My stubborn realism gave rise to a lively argument, and a great battle ensued. Though later an armistice was called, and neither could consider himself the victor, yet each considered himself invincible. Sentences like the following made me quite unhappy: "How could any experience ever be gauged by an idea, for the characteristic thing about an idea is that it can never be congruous with an experience." Yet if he termed an idea what I called an experience, then there must certainly be something negotiable, something in common between us. The first step had after all been taken, and Schiller had great personal magnetism and the power to hold those whom he had attracted. Thus I became interested in his plans and promised to turn over to *The Hours* some of the manuscripts hidden away in my desk. His wife, whom I had known and loved from her childhood, contributed her share to an enduring friendship, all our mutual friends were happy; and thus through the great duel between the objective and the subjective, we sealed a bond which lasted uninterruptedly and accomplished much good for ourselves and others.

After this happy beginning, during ten years[11] of intimate association, such philosophic tendencies as were latent in my nature gradually unfolded. It is my intention to account for this unfolding insofar as it is possible, but the difficulties involved must be strikingly evident to the initiated. Those who command a higher vantage ground from which to survey the easy confidence of the human mind, the understanding innate in the healthy individual with no doubts about facts and their significance nor about his own ability to comprehend, judge, and properly evaluate them—men who command such a vantage point will freely admit that a virtual impossibility is undertaken when one attempts to describe transitions to a purer, freer, and more objective phase—transitions of which there must be thousands upon thousands. We are not speaking here of levels of education, but of those false trails, pitfalls, and circuitous bypaths which are followed by sudden progress and a vigorous upswing to a higher state of culture.

What individual can actually assert that scientifically he always travels in that highest area of consciousness, where external things can be observed with utmost deliberation, with impartiality and vigilance alike, where he works in accord with the law of his inner nature, with vision and foresight and with the enduring hope of acquiring a truly pure and harmonious point of view. Does not the world, do we not ourselves tarnish

---

[11] Schiller died in 1805.

the lustre of such moments? But at least we may cherish pious hopes, and a loving attempt to reach the unattainable is not denied us.

Whatever victories we achieve in the venture, we shall commend to long-esteemed friends and to the youth of Germany in their striving toward the good and the right.

May we thus attract and acquire vigorous sympathizers and future exponents!

Indecision and Surrender[12]

IN OBSERVING the cosmic structure from its broadest expanse down to its minutest parts, we cannot escape the impression that underlying the whole is the idea that God is operative in Nature and Nature in God, from eternity to eternity. Intuition, observation, and contemplation lead us closer to these mysteries. We are presumptuous and venture ideas of our own; turning more modest, we merely form concepts that might be analogous to those primordial beginnings.

At this point we encounter a characteristic difficulty—one of which we are not always conscious—namely, that a definite chasm appears to be fixed between idea and experience. Our efforts to overbridge the chasm are forever in vain, but nevertheless we strive eternally to overcome this hiatus with reason, intellect, imagination, faith, emotion, illusion, or—if we are capable of nothing better—with folly.

By honest persistent effort we finally discover that the philosopher might probably be right who asserts that no idea can completely coincide with experience, nevertheless admitting that idea and experience are analogous, indeed must be so.

In all scientific research the difficulty of uniting idea and experience appears to be a great obstacle, for an idea is independent of time and place but research must be restricted within them. Therefore, in an idea, the simultaneous and successive are intimately bound up together, whereas in an experience they are always separated. Our attempt to imagine an operation of Nature as both simultaneous and successive, as we must in an idea, seems to drive us to the verge of insanity. The intellect cannot picture united what the senses present to it separately, and thus the duel between the perceived and the ideated remains forever unsolved.

For this reason we justifiably take flight into poetry, giving by way of change a new form to an old song:

---

[12] Written in 1818; published in *Natural Science in General; Morphology in Particular,* Vol. I, No. 1 (1820).

> Regard with silent wonder
> The Eternal Weaver's masterpiece.
> A single movement sends the shuttle
> Over, under, till the myriad threads
> Meet and interlace, creating
> Countless unions at one stroke!
> The warp, not mounted thread by thread,
> But laid down in the timeless past
> Awaits the casting of the weft,
> Forever waits the Master's will.
>
> —GOETHE: *Antepirrhema*
> (English rendering by Aldyth Morris)

THE OBJECTIVE AND SUBJECTIVE RECONCILED BY MEANS OF THE EXPERIMENT[13]

AS SOON AS an individual becomes aware of the things around him, he looks upon them with reference to himself, and quite rightly so, for his whole fate depends on whether they please or displease him, attract or repel, benefit or harm. Convenient and necessary as this natural way of regarding and judging things appears to be, an individual is subjected to a thousand errors in its application—errors that often confuse and embitter him.

A far more difficult labor is undertaken when an individual with a vigorous urge toward knowledge desires to observe the things of Nature in themselves and in their interrelations. He soon feels the lack of the measuring rod so helpful to him in examining things with personal reference to himself. Now he is deprived of the gauge of pleasure and displeasure, attraction and repulsion, benefit and harm. He must renounce such a standard completely; he must be disinterested and almost godlike in his seeking; he must investigate what exists, not what gives him pleasure. Thus, the true botanist must not allow himself to be influenced by either the beauty or the utility of plants; his duty is to study their formation, their relationship to the rest of the plant world; and like the sun that lures forth and illuminates each and every plant, the botanist too must observe and survey them all with the same calm glance, deriving his standard for recognising and judging them not from within himself but from the area of the things themselves.

---

[13] Written in 1792–1793; published in *Natural Science in General; Morphology in Particular,* Vol. II, No. 1 (1823).

As soon as we observe a thing with reference to itself and in relation to other things, forswearing personal desire or aversion, we shall be able to regard it with calm attention and form quite a clear concept of its parts and its relationships. The further we continue these observations, the more we are able to provide links between isolated things, and the more we are able to exert our powers of observation. If we can manage to apply such knowledge in our daily affairs, we are deservedly deemed intelligent; yet to be intelligent is not difficult for a well-organized individual, one moderate by nature or made so through experience, for life puts us right with every stride we take. However, if the observer is to apply this same sharp power of judgment in examining the mysterious relations of Nature, if he is to heed his steps in a sphere in which he is almost alone, if he is to guard against all premature conclusions, if he is to have his goal ever in mind without allowing any circumstance along the way, be it advantageous or disadvantageous to his theory, to go unnoticed; if he is to be his own severest critic even in situations where he cannot easily be disproved by others; if he can be distrustful of himself even in his most zealous efforts: the severity of the demands made upon him are unmistakable and likewise the impossibility of their complete fulfillment, regardless of whether the demands are made by himself or others. And yet these difficulties, one may undoubtedly say this hypothetical impossibility, must not deter us from doing the utmost within our power. We shall at least make greater progress if we attempt to visualize the means through which outstanding men have been able to advance the sciences, if we clearly point out the false trails upon which they have gone astray, followed by a great number of pupils, often over a period of centuries, until later experiences have once again directed the observer to the right path.

No one will deny that experience has and should have the greatest influence in natural science, of which I am speaking here primarily, as it does in every other undertaking. Indeed, one would as little deny the lofty and almost independent creative powers of the mind in which these experiences are comprehended, collected, arranged, and developed. However, the manner of undergoing and applying these experiences, of developing and using our faculties, cannot be as generally known or recognized.

As soon as one directs the attention of alert, astute individuals to certain phenomena, one finds that they have both predilection for and skill in observation. I have often noticed this fact in my zealous study of the science of light and color, since I frequently discuss the subject of my current interest with persons ordinarily not accustomed to such observations. As soon as their interest is stimulated, they notice phenomena with which I in part was unacquainted and in part had overlooked. In that way they rectify my prematurely formulated ideas, thus giving me the opportunity to advance more rapidly and to emerge from the limitations beleaguering one during a laborious investigation.

Here we find what is true of so many human undertakings, namely, that it is only the interest of several people focused upon one subject that enables us to produce outstanding results. Thus it is apparent that the greatest obstacle to the scientist may be the envy prompting him to exclude others from the honor of a discovery, or an immoderate desire to treat and work out a discovery exclusively in his own way.

In the past I have enjoyed working with others far too much, not to wish to continue in the same way. I know precisely to whom I am indebted for this or for that in my career and it is my intention to admit it openly and joyfully in time to come.

Now if lay individuals of native alertness can be of such great help to us, how much more extensive must be the benefit when experts work hand in hand! For one thing, each science in itself is such a massive thing that it will occupy many individuals, and cannot be mastered by one individual alone. We may point out that knowledge is like flowing water confined by a dam and gradually lifted to a higher level, for the most magnificent discoveries are made not so much by individuals as by an age. Evidence of this are the many important discoveries made simultaneously by two or even more trained thinkers.

On the one hand we are in great debt to society and friends; on the other hand we owe an even greater debt to the world and the century. In either case it is impossible to exaggerate the necessity for sharing of ideas, for co-operation, criticism, and opposition, if we wish to keep to the right path and to forge ahead.

Thus in scientific pursuits we must do the reverse of what the artist finds advisable. The latter does right not to allow his work of art to be viewed until it is finished, for it is difficult to advise him or give him assistance. Yet when his work of art is finished, he must then consider and take to heart the praise or the censure, in that way preparing himself for a new piece of work. But in things scientific it is beneficial to share each individual experience with others, and even each supposition; and it is highly advisable not to erect a scientific structure before the blueprints and materials are generally known, evaluated, and chosen.

We give the term experiment to the process of systematically repeating the experiences of predecessors, contemporaries, or ourselves, and of reproducing phenomena that have arisen in part by chance, in part by plan.

The value of an experiment consists primarily in the fact that, simple or complex, it can be repeated at any time with certain apparatus and with the required skill, as often as all the prerequisite conditions can be united. We justly admire the human intellect when we observe the ingenious combinations devised for such purposes and the paraphernalia that continue to be invented daily.

Meritorious as each individual experiment may be, it nevertheless can have value only in combination and in connection with others; however, to combine and connect two somewhat similar experiments, demands more strictness and discernment than even astute observers ordinarily require of themselves. Two phenomena may be allied to one another, but by no means so closely as we believe. Two experiments can appear to follow one from another, but a long series of links would still have to be discovered between them before they could be brought into a truly natural relationship.

One cannot guard too closely against drawing premature conclusions: for it is in the transition from experiment to conclusion, from knowledge to application, that all one's inner enemies lie in wait as at a mountain pass—imagination, impatience, rashness, self-complacency, rigidity, conventionality, prejudice, sloth, frivolity, fickleness, and all the rest, all lurk in ambush here to surprise and overpower the active man of the world, and even the quiet retiring observer usually thought to be safeguarded against all violent emotions.

To warn against this danger, greater and more imminent than one thinks, and to arouse greater vigilance, I should like to set up a series of paradoxes. I venture to assert that *one* experiment, even several experiments combined, prove nothing; indeed, that nothing can be more dangerous than the attempt to confirm a theory by experiments; and that the greatest errors have arisen precisely because its dangers and its inadequacies were not realized. However, I must be even more explicit, lest I be accused of merely wanting to say something spectacular.

Every single experience we have, each experiment whereby we repeat it, is actually an isolated part of our knowledge; through frequent repetitions we verify these isolated data. Two experiments in the same field, though indeed closely related, will appear even more so if they come to our knowledge simultaneously, and we will then be inclined to consider them more closely related than they actually are. This is in keeping with human nature; the history of the human intellect provides us with thousands of examples; I myself have often noticed my own tendency to commit this error.

The mistake is closely related to another, one indeed from which it usually arises, namely, that we take more delight in our concept of a thing than in the thing itself, or more correctly, we have joy in a thing only insofar as we form an idea about it. It must fit in with our way of thinking, and though we elevate and purify our mode of thought ever so much above the common level, in most cases it remains a mere attempt to bring sundry things into a comprehensible relationship, which, strictly considered, does not exist. This is the explanation for our inclination to hypotheses, terminologies, and systems, an inclination which we cannot after all condemn, as it arises logically from the organization of our nature.

If on the one hand each individual experiment must be regarded as isolated by its nature, and if on the other the strength of the human intellect is powerfully exerted toward integration of all exterior things coming to its attention, we may fail to realize the danger involved in attempting to combine a single experiment with our preconceived idea or to prove by means of experiments a relationship which is not actually sensory, yet has already been created through the power of the intellect.

Often theories and systems resulting from such efforts find greater approval than they merit and are maintained longer than is warranted. In such instances, they retard and weaken the advance of the human intellect, though admittedly fostering it in some respects. For it often happens that the fewer the data available to a man of good intellect, the greater the ingenuity he will use in interpreting them. To show his mastery, as it were, he will select from the data a few favorites that flatter him, he will manage to arrange the rest so that they will not appear to contradict him, and lastly he will complicate, obscure, and eliminate the hostile data. Thus in the end, the whole no longer resembles a free republic but a despotic court circle.

A man of such great ingenuity will have no lack of advocates and pupils to become acquainted with this fabrication through book learning and to admiringly assimilate as much as possible of their master's mode of thought. In this way a theory will often gain the whiphand to the extent that an individual is considered bold and impertinent if he so much as dare express a doubt. Not until succeeding centuries will anyone venture to attack the sacrosanct, to show greater common sense in treating the subject, to resubmit the data to a more normal way of thinking, to take the theory less seriously, and to say about the founder of this scientific sect what a wit once said about a great natural philosopher, namely, that he might have been a great man had he been less ingenious.

It may not be sufficient to point out and warn against the danger. It is only right to speak out and make known whether one has himself avoided such false trails and whether predecessors have done so.

I have just remarked that I consider dangerous the direct application of an experiment to prove a given theory, and have thereby indicated that I consider an indirect application advantageous; and since everything revolves about this point, I consider it essential to be more explicit.

In organic life nothing is unconnected with the whole, and even if the phenomena *appear* isolated to us, even if it is necessary to regard experiments as mere isolated facts, it does not prove that they actually *are* isolated. The question is merely how to find the interrelation of these phenomena, of these occurrences.

We have seen above that the individuals most prone to error are those who immediately interpret an isolated part on the basis of their preconceptions. Conversely, we find that the greatest success is achieved by those who do not cease to investigate and work out all possible aspects and variations of a single experiment.

Since the more general forces and elements of Nature especially are in eternal action and reaction, one can say that all phenomena are connected with countless others. For instance, we say of a free-moving point of light that it sends its rays in all directions. Therefore, when we have evolved such an experiment, it is impossible to be too careful in investigating what borders and what follows directly upon it. Indeed, we must even pay greater heed to this than to things referring directly to the current experiment itself. The real duty of the scientist is thus to modify each and every single experiment, this duty being just the opposite of the writer's duty to entertain. The writer bores his readers if he leaves nothing to the imagination; the scientist must work indefatigably, as if he desired to leave nothing for his successors to do—even though the disproportion of the human intellect to the nature of the universe must all too soon remind him that no man has the capacity for making a definitive presentation of any subject whatsoever.

In my first two contributions to optics, I have attempted to establish a series of experiments which directly adjoin and touch each other, which, if one is familiar with all of them and sees them in proper perspective, merely represent manifold aspects of the same experiment.

An experiment consisting of several individual experiments is obviously of a higher type. It presents a formula by which numerous single calculations are expressed. I consider the highest duty of the scientist to work forward toward experiments of this higher type, inspired by the example of the outstanding men who have worked in this field.

Taking an example from the mathematician, we ought to learn the wisdom of proceeding step by step, or rather of deriving one fact from the preceding one; and even when we are using no mathematical calculations, we ought always to set to work as though we were under obligation to give an accounting to the strictest geometer. And it is indeed the mathematical method, because of its exactness and purity, that immediately reveals possible flaws in an assertion; its proofs are actually detailed demonstrations that everything brought forth in association has already been present in its single parts and in its entire sequence, has been surveyed in its entirety and found to be correct and incontrovertible under all conditions. Thus mathematical proofs are always *expositions* and *recapitulations*, never *mere arguments*. Having mentioned this distinction, kindly permit me to present it in somewhat greater detail.

It is easy to see the great difference between mathematical demonstrations, which carry primary elements through various combinations, and evidence which clever debaters can derive from arguments. Through wit and imagination, arguments containing extremely isolated factors can be brought to a focal point with surprising ease, producing the illusion of right or wrong, true or false. In the same way, a more or less misleading conclusion can be reached by linking individual experiments together for the sake of an hypothesis.

However, the individual desiring to set about his work with honesty to himself and others, will perform the individual experiments with utmost care and thus develop data of a high level. These data, placed side by side in concise and easily understood propositions, can then be arranged in the same order in which they developed one by one and can be brought into a logical relationship so that they stand unshakable, singly and collectively, like mathematical formulae.

The elements of such a superior type of experiment, which actually is made up of many individual experiments, can thus be examined and tested by anyone, and it is not hard to judge the possibility of expressing the many single parts by a general proposition. For here nothing arbitrary takes place.

However, with the other method, in which we attempt to prove an assertion by isolated experiments, by arguments as it were, the conclusion is often arrived at by trickery or it may be entirely unconvincing.

When a series of experiments of the higher type have been brought together, we may apply reason, imagination, and wit to any extent desired, and no harm will be done, indeed it will be advantageous. We cannot possibly be careful, diligent, or even pedantic enough in performing the first step, inasmuch as it is undertaken for posterity as well as for the present time. But these materials must be arranged and set down in natural series, not put together hypothetically, not arbitrarily systematized. Then any individual will be free to combine them in his own way, to form a whole more or less convenient and pleasing to human conception in general. In this way we can keep separate what should be kept separate and increase the collected data much more rapidly and correctly than if we have to put later experiments aside unused, like stones that have been hauled in after the structure is already complete.

The opinions and precepts of outstanding men encourage the hope that I am on the right path, and with this explanation I expect to satisfy friends who sometimes inquire about the "real" purpose of my optical studies. My purpose is this: to collect all data in the field, to set up my own experiments and carry them out in the greatest diversity, by methods easily duplicable and within range of more individuals than heretofore; furthermore, to formulate the propositions by which data of the higher type can be expressed and to have the patience to learn whether these too may be subordinated under a higher law. If imagination and wit should nevertheless impatiently hurry ahead, the procedural method will itself indicate the point to which they must again return.

Experience and Science[14]

PHENOMENA, also called facts in lay language, are certain and definite by nature, but often indefinite and variable as they meet the eye. The scientist attempts to grasp and hold fast what is definite in the phenomena; in individual cases he is concerned not only with their actual but also with their ideal appearance. As I have occasion to notice in my present field of work, empirical breaks must often be disregarded in order to preserve a pure, constant phenomenon. However, as soon as I permit myself to do this, I am establishing a kind of ideal.

Nevertheless, a vast difference exists between disregarding whole sequences in favor of a hypothesis, as theorists often do, and the sacrifice of a single empirical break in the interest of preserving the idea of the pure phenomenon.

Since, therefore, the observer never sees pure phenomena with his eyes, since much depends instead upon his own state of mind, on the state of the organ itself at the moment, on light, air, weather, bodies, treatment, and a thousand other things, it would be like attempting to drink up the ocean if he were to fasten upon each and every phenomenon with the intention of observing, measuring, judging, and describing them individually.

---

[14] Written in January, 1798; sent to Schiller; first published in the Weimar edition. Its contents directly supplement the foregoing essay.

In my observation and contemplation of Nature, especially of late, I have remained as faithful as possible to the following method.

After observing a certain degree of constancy and logical sequence in phenomena, I derive an empirical law and prescribe it for future phenomena. If the law and the phenomena later coincide completely, I am vindicated; if they do not, at least my attention has been drawn to the details of individual cases and to the necessity for more correct methods of organizing the contradictory experiments. However, if under similar circumstances, a case appears that contradicts my law, I know I must pass on and search for a higher viewpoint.

According to my experience, this then would be the point where the human mind could approach closest to the objects in their general aspects, absorb them, and in rational terms amalgamate with them (as we also do at the practical level).

The object of our work would then be to demonstrate: (1) the *empirical phenomenon,* of which every individual is conscious in Nature and which later is elevated to (2) a *scientific phenomenon* by experimentation, by representing it under circumstances and conditions differing from those in which we first encountered it, and in a more or less effective sequence; and (3) the *pure phenomenon* now standing forth as the result of all experiences and experiments. It can never be isolated, appearing as it does in a constant succession of forms. In order to describe it, the human intellect determines the empirically variable, excludes the accidental, separates the impure, unravels the tangled, and even discovers the unknown.

Here we would reach the ultimate goal of our powers, if the human being knew his place. For we are not seeking causes but the circumstances under which the phenomenon occurs. Its logical sequence, its eternal return under a thousand conditions, its uniformity and mutability are considered and accepted; its definiteness is recognized and redefined by the human intellect. And in my opinion such work is certainly not mere speculation, but rather the practical and self-correcting operations of ordinary common sense as it ventures out into a higher sphere.

INFLUENCE OF THE NEW PHILOSOPHY[15]

FOR PHILOSOPHY in the narrower sense I had no aptitude. It was only the continuous counterinfluence I was forced to exert against the inrushing

---

[15] Written in September, 1817, printed in *Natural Science in General; Morphology in Particular,* Vol. I, No. 2 (1820).

world, in my effort to win the mastery over it, that led me to examine various philosophical views. These I sought to grasp almost as though they were objects, and through them I sought to cultivate my mind. Brucker's *History of Philosophy* I had read assiduously in my youth, but as regards philosophy itself I had always resembled those individuals who have seen the starry sky overhead all their lives and can distinguish a few striking constellations without understanding anything about astronomy, who know the Great Bear but not the polar star.

Art and its theoretical requirements I had often discussed with Moritz[16] in Rome, and a small publication still bears witness to our fruitful innocence at that period. Furthermore, the work in plant metamorphosis was bound to result in a natural method; for as vegetation demonstrated its method to me step by step, I could not possibly go astray, and by letting vegetation tell its own story, I necessarily became acquainted with its ways and means of gradually developing to completion the most obscure phenomena.

In my study of physics I gained the conviction that one's highest duty in observing phenomena is to trace accurately every condition under which a phenomenon makes its appearance and to aim at observation of as many phenomena as possible. To an observant investigator, inasmuch as they follow one upon the other and indeed overlap, each virtually reveals its organization and manifests its life course. Meanwhile this was a twilight state; nowhere did I find the clarification I desired; and in the last analysis an individual can be enlightened only by methods in harmony with his own disposition.

Kant's *Critique of Pure Reason* had long since appeared,[17] but it lay completely beyond my orbit. Nevertheless, I was present at many a discussion of it, and with some attentiveness I could notice the old question continually reappearing, namely, how much we ourselves and how much the outside world contributes to our intellectual existence. I had never separated the two and when I did philosophize about subjects in my own way, I did so with unconscious naiveté, in the belief that I actually saw my views before my very eyes. Whenever the familiar dispute arose I was inclined to align myself on the side that did greatest honor to the human race. Thus I fully agreed with those of my friends who, with Kant, maintained that even though all our knowledge concerns experience, it does not necessarily arise from experience. Knowledge *a priori* I could grant quite as much as synthetic judgment *a priori*, for throughout my life, both as poet and scientist, I had proceeded

---
[16] See p. 216.
[17] In 1781.

analytically and synthetically by turns. The systole and diastole of the human intellect had in my eyes never appeared separated but pulsative instead, like breathing. For all this I had no words, and least of all fine phrases, but for the first time in my life a philosophical theory had actually appealed to me. However, it was merely the portal that pleased me and I did not trust myself to venture into the labyrinth itself. Sometimes it was my poetic gift that created the obstacle, at other times my understanding. Thus I felt in no wise improved.

Unfortunately, Herder was a student of Kant and yet his opponent. I myself was even worse off; for I could neither agree with Herder nor follow Kant. Meanwhile I continued a serious investigation of the formation and transformation of organic bodies, and found that my method of treating plants was reliable as a guide. It did not escape me that Nature, although always observing an analytic procedure, developing from one vital mysterious whole, sometimes appeared to operate synthetically, since elements that seemed to be completely foreign were juxtaposed and joined together. Again and again, then, I returned to the Kantian theory, feeling I understood some chapters better than others and obtaining much that I could use in my own work.

Then, I chanced upon the *Critique of Judgment,* a work to which I owe an extremely happy period in my life. Here I found my two most disparate interests juxtaposed; the results of both art and science were discussed, and aesthetic and teleological judgments were mutually clarified.

Even though it was not always possible to reconcile my conception with that of the author, even though here and there something seemed to be lacking, nevertheless the chief thoughts of the great work were quite analogous to my literary work, to my way of life, and to my method of thinking. The inner life of Art and Nature, their mutual actions directed outwardly from within, were clearly expressed in the book. The products of these two endless infinite worlds were explained to exist, each for its own sake, and whatever stood side by side, though evidently existing *for* one another, did not exist expressly *because* of one another.

My aversion toward final causes was now explained and vindicated; I could clearly distinguish between purpose and effect, and also understand why the human mind sometimes interchanges them. It pleased me that poetry and comparative natural science were closely related, subject to the same standard of judgment. Aroused to enthusiasm, I made faster progress precisely because I did not know whither my methods were leading me and because I won little approval from the Kantians themselves of what I was accomplishing and how I was doing it. Thrown back upon myself, I studied the book repeatedly at various intervals. I still take delight in the underlined passages in my old copy, for instance, those in the *Critique of Pure Reason,* into which I now succeeded in delving deeper and deeper, inasmuch as the two works, arising from one intellect, helped explain each other. I did not succeed to the same extent with the Kantians themselves; they listened to me, of course, but could give me nothing in return nor in any way help me to advance. More than once I had the experience of hearing one of them confess to me with smiling perplexity that my method was an analogue of the Kantian concept, a strange one however.

What an odd thing it really was, did not emerge until my acquaintance with Schiller had developed into friendship. Our conversations were theoretical or thoroughly practical, usually both at once: he preached the gospel of freedom, I refused to see Nature's rights curtailed. More perhaps through friendship than from conviction, he avoided those hard expressions in his allusions to the Good Mother that had made his essay "On Grace and Dignity"[18] so odious to me. I on my part, obstinate and headstrong, was not content with stressing the advantages of the Greek method of poetic creation and the poetry originating therefrom, but kept insisting that this type of poetry was exclusively and solely the correct and desirable one. In this way he was compelled to more incisive thought, and it is to this very conflict that we owe his essays on sentimental and naive poetry.[19] The two methods of writing here yielded to each other and, though opposites, granted each other equal rank.

In this way Schiller laid the groundwork for a whole new system of aesthetics; the terms *Hellenic* and *Romantic,* and whatever else might be found by way of synonyms, can all be traced back to those early discussions of the superiority of realistic or idealistic treatment.

Thus, little by little, I became accustomed to a language which at first had been completely foreign to me. Now my progress was easier, because the new language favored a higher concept of art and science. I felt enriched and ennobled, whereas formerly we others had been made to feel unworthy by the treatment we received from popular philosophers—and from another type of philosopher for whom I am hard pressed to find a name.

---

[18] "On Grace and Dignity in Literature," 1793. See pp. 216, 217.
[19] Schiller's famous essay, "On Naive and Sentimental Poetry," 1795, in which Goethe was described as the "naive" or classic poet, organically one with nature, and Schiller himself as the "sentimental" poet, who had lost this elemental tie with nature.

Further progress I owe particularly to Niethammer,[20] who patiently and affably endeavored to unseal the chief mysteries for me, to unfold and to explain individual concepts and expressions. What I owed at the time, and later, to Fichte, Schilling, Hegel, the brothers Humboldt, and the brothers Schlegel, I may possibly account for at some later time. May it still be vouchsafed me to sketch, if not fully describe, my attitude toward an epoch which was so important for me, namely, the last decade of the past century.

Intuitive Judgment[21]

In my efforts to utilize if not actually master the Kantian theory, it sometimes seemed to me as if the worthy man were proceeding roguishly and ironically, at one point appearing to set narrow limits for our perceptive capacity and at another beckoning us furtively beyond them. To be sure, he may have noticed how arrogantly and cockily a man proceeds when, unencumbered by much experience, he straightway and unthinkingly rejects one consideration in a premature endeavor to establish another, or links up the subjects under study with some caprice or other flitting through his mind at the moment. Our master thus restricts his thinkers to reflective and expository judgment, sternly forbidding determinative judgment; but after driving us sufficiently into a corner and even bringing us to despair, he suddenly decides in favor of the most liberal interpretations and allows us to make what use we will of the freedom he has in some measure vouchsafed us.

In this connection the following passage was highly significant to me:

"In contrast to our own analytical intellect, we can conceive of an intuitive one which proceeds from the synthetically universal (the concept of the whole as such) and advances to the particulars, in other words, advances from the whole into its parts. At this point it is not necessary to prove that such an *intellectus archetypus* is possible, but merely that we are inevitably led to it when we contrast our own analytical, image-requiring intellect *(intellectus ectypus)* with its own fortuitous character, and that the idea of an *intellectus archetypus* would contain no contradiction."[22]

---

[20] Friedrich Immanuel Niethammer, 1766–1848, one time professor of philosophy, Jena.

[21] Written in 1817, this essay appeared in *Natural Science in General; Morphology in Particular*, Vol. I, No. 2 (1820).

[22] From *Critique of Judgment*, par. 77.

To be sure, the author seems to be referring here to godlike understanding; yet since it is possible in the moral realm to ascend to a higher plane, drawing close to the Supreme Being through faith in God, virtue, and immortality, the same might well hold true in the intellectual realm. Through contemplation of ever-creative Nature we might make ourselves worthy of participating intellectually in her productions. Had not I myself ceaselessly pressed forward to the archetype, though at first unconsciously, from an inner urge; had I not even succeeded in evolving a method in harmony with Nature? What then was to prevent me from courageously embarking upon the adventure of reason, as the old gentleman of Königsberg himself calls it?

The Creative Urge[23]

IN HIS *Critique of Judgment* Kant takes the following stand on the subject mentioned in our title: "In regard to the theory of epigenesis,[24] no one has contributed more toward its proof, toward discovery of effective principles for its employment, toward guarding against its excessive application, than has Herr Blumenbach."[25]

Such testimony from the conscientious Kant induced me to again take up Blumenbach's book, which I had read but not mastered. Here I found my Caspar Friedrich Wolff[26] occupying a position midway between Haller and Bonnet on the one side and Blumenbach on the other. Wolff, in behalf of his epigenesis, was compelled to presuppose an organic element by means of which the creatures destined for organic life are nourished. He assigned to this matter a *vim essentialem*, which, incorporating itself with everything about to reproduce, lifts itself to the rank of a producer.

To me, expressions of that sort leave something to be desired; for no matter how dynamic one visualizes the substance, something material still adheres to it. The word "force" likewise designates primarily something merely physical, indeed even mechanical, and what is to be organized from the organic substance remains a dark, obscure point.

---

[23] Written 1817–1818; printed in *Natural Science in General; Morphology in Particular*, Vol. I, No. 2 (1820).

[24] Genuine development of newly arisen plant and animal individuals under the effect of forces always present and always operating anew, as contrasted with the unfolding of finished entities in the Bonnet-Haller preformation theory.

[25] Johann Friedrich Blumenbach, 1752–1840, anthropologist, Göttingen. See p. 121.

[26] See p. 176.

Then Blumenbach achieved the highest, the ultimate in expression by personifying the mystery, calling the phenomenon we are considering a *nisus formativus,* a creative urge, a vigorous activity effecting formation.

Penetrating more deeply into all this, we would have an easier, briefer, and perhaps more fundamental task if we admitted to ourselves that, in order to study existent phenomena, we have to presuppose a previous activity, that we have to give that activity an appropriate element as a basis of operation, and finally that we have to imagine the activity to be continually existent and eternally available. Personified, it becomes a god, a creator and preserver, whom we feel compelled to worship and extol in all possible ways.

If we return to the field of philosophy to reconsider evolution[27] and epigenesis, these terms appear to be words that merely beg the question. Especially the insertion theory soon repels the discriminating; with the theory of absorption and assumption, something absorbed and assumed must always be presupposed; and if we are disinclined toward the idea of preformation, we nevertheless resort to predelineation, predetermination, pre-establishment, and whatever other terms may exist to describe what is antecedent to perception.

This much, however, I venture to say: in considering an organic entity, unity and freedom of the creative urge are incomprehensible without the concept of metamorphosis.

In closing, let me offer an outline to stimulate further thought:

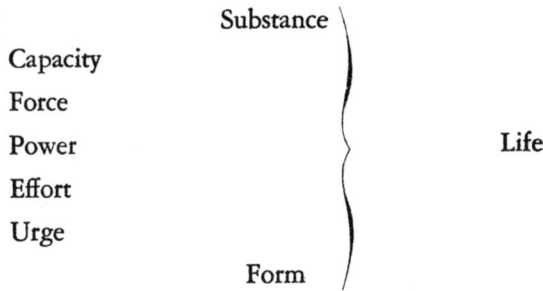

---

[27] Used in the old sense of the term, synonymous with preformation.

Considerable Assistance from One Ingeniously Chosen Word[28]

IN DR. HEINROTH'S[29] book on anthropology, a work to which we shall return repeatedly, the author speaks favorably of my character and work. Indeed, he calls my method of procedure unusual, saying that my capacity for thinking is *objectively* active. By this he means that my thinking is never divorced from objects, that the elements of the objects and my observation of them interpenetrate, become fused in the process of thought; that my observation is itself thinking, and my thinking is a way of observation; and this is a method to which my friend Heinroth gives his approval.

The following pages will serve to indicate how stimulating I found every single word, accompanied by such praise. I commend these pages to the interested reader, after he has previously acquainted himself with the details of the subject on page 389 of the book cited.

In the present issues I have endeavored, as formerly, to describe my way of regarding Nature, and, at the same time, to reveal as far as possible my cast of mind, my way of being. For this purpose an older essay, "The Objective and the Subjective Reconciled by Means of the Experiment,"[30] will prove helpful.

Here I confess that the great and important-sounding advice, "Know thyself," always appeared to me to be open to question, as a ruse of conspiring priests intent upon confusing the laity with unattainable ideals, upon seducing them from active life to dangerous introspection. Man knows himself only insofar as he knows the world, becoming aware of it only within himself, and of himself only within it. Each new subject, well observed, opens up within us a new vehicle of thought.

Most helpful of all are our fellow men, since they have the advantage of judging us and our environment from their own points of view, and may therefore achieve more precise knowledge of our mentality than we ourselves are capable of attaining.

I have therefore in my more mature years paid much attention to the analyses others have made of me, in order that in and through them, as though reflected in so many mirrors, I may better understand myself and my thinking.

---

[28] Written in the spring of 1823; first printed in *Natural Science in General; Morphology in Particular*, Vol. II, No. 1 (1823).
[29] Johann Christian Heinroth, 1773–1843, professor of psychiatry, Leipzig. His *Textbook of Anthropology*, with the passage on Goethe, had appeared in 1822.
[30] See pp. 220–227.

Opponents do not come into the picture. My existence is odious to them; they reject the goals toward which my activity is directed and consider the instruments for achieving these goals as so much wasted effort. I therefore dismiss and ignore these opponents; they cannot contribute to my advancement, and it is advancement on which everything in life depends. However, where my friends are concerned, I am just as happy to be curbed as to be directed toward infinity. I always give heed to their opinions, confident that they will contribute to my development.

What I have just said about objective thinking, I may likewise say of objective writing. So deeply did certain motifs, legends, ancient history impress my mind that I kept them in my mind for forty to fifty years, alive and at work. To me it seemed a most wonderful boon that such valuable images could be renewed in one's imagination, transformed constantly, without change, maturing into a purer state and into clearer form. I should like to mention here only "The Bride of Corinth," "The God and the Bayadere," "The Count and the Dwarves," "The Singer and the Children," and finally "Paria," soon to be published.

What has been said above also explains my inclination toward the occasional poems to which I was irresistibly inspired by special occurrences. And thus it has been noticed that something particular is always at the basis of each of my poems, a definite nucleus for a more or less significant product; and for this reason the poems have not been sung for some years. This is especially true with those of definite character requiring the singer to transfer himself from a generally indifferent state to a special, strange point of view and atmosphere, and to articulate the words clearly, so that one will know what the song is actually about. On the other hand, stanzas of nostalgic content find readier approval, and mine have achieved some popularity, along with other German products of the same variety.

Directly linked to this consideration, is the tenor of my opinion on the French Revolution, over a period of many years, and my boundless effort to poetically master this most terrible of all events, to trace its causes and results. If I look back upon the years, I clearly see that my attachment to this unsurveyable subject has long, almost unprofitably, been devouring my ability as a poet; yet that impression has taken such deep root with me that I cannot deny that I still am pondering a continuation of *The Natural Daughter,* that I am still developing this wonderful production in my thoughts without having the courage to devote myself in detail to its execution.

Turning now to the *objective thinking* attributed to me, I find I was compelled to follow the same method in natural history too. What trains of thought and reflection were indeed necessary before the idea of plant metamorphosis could take form in my mind, as revealed to my friends in my *Italian Journey*.

It was likewise with the concept that the skull consists of vertebrae. The three hindmost I soon recognized, but not until the year 1791, when I picked up a battered sheep's skull from the sand of the dunelike Jewish cemetery in Venice, did I realize instantaneously that the facial bones likewise could be interpreted as originating from vertebrae. I clearly saw the transition from the first wing-bone to the ethmoid and the conchae, and thus had a view of the whole in its most general aspects. This brief account may be sufficient to indicate what was achieved at that time, and now I should still like to point out briefly how that expression "objective thinking," coined by a generous and intelligent man, is of help to me at present.

For several years already I have been attempting to revise my geognostic studies, especially with the view of seeing to what extent I might reconcile the studies and the convictions they gave rise to, with the new volcanic theory spreading everywhere. Up until now reconcilement seemed impossible. However, I was suddenly enlightened by the word "objective," clearly realizing that all the objects I had been observing and examining for fifty years were bound to arouse in me precisely the concept and the conviction from which I cannot now depart. To be sure, I can for short periods shift to that other viewpoint, but I must always return to my old method of thinking if I am to feel at all comfortable.

Inspired by these considerations, I continued to examine myself and discovered that my whole procedure rests upon deduction. Because I am cautious and faithful in my work—both in what I accept from others and in what I achieve by myself—I do not rest until I find a pregnant idea from which much can be deduced, or rather a point which voluntarily yields and brings many things to me. When a phenomenon appears that I cannot solve by deduction, I allow it to remain unsolved, as a problem, and this method of procedure I have found advantageous throughout my long life. Even though it is impossible at first to unravel the origin and the interconnections of a problem, and I set it aside for a long time, suddenly, after years have passed, I find everything clarified, coherently outlined. I shall therefore take the liberty of continuing in these pages my historical presentations of previous experiences and observations, and the manner of thinking arising therefrom. In that way I can at least achieve a characteristic credo for myself, give insight to my opponents, assistance to individuals sharing my views, instruction to posterity, and if at all possible, a contribution toward harmonious relations in general.

Analysis and Synthesis[31]

In his third lecture this year, Herr Victor Cousin[32] has commended the eighteenth century principally for devoting itself to analysis in the sciences and for being on its guard against premature synthesis, in other words, against hypotheses. Nevertheless, after advocating the analytic method almost exclusively, he finally concedes that we cannot avoid synthesis completely and must resort to it from time to time, with caution.

In considering these remarks, our first reaction was that even in respect to synthesis the nineteenth century still has a great deal left to do; friends and experts of science ought to recognize our failure to analyze, develop, and clarify the false syntheses transmitted to us, and to restore to the intellect its old privilege of taking a direct view of Nature.

Let us mention here by name two such false hypotheses: the decomposition of light and its polarization. Both are mere names, conveying nothing to the thinking individual and yet so often repeated by scientists.

In our observation of Nature, it is not enough to apply the analytic process, that is, to reduce the subject to as many details as possible and in this manner become acquainted with them. We must also apply the same analytical method to the existing hypotheses, to see whether the original procedures were correct and in harmony with genuine method.

It is for this reason I have made a detailed analysis of Newton's work. His was the mistake of constructing a hypothesis on the basis of a single phenomenon—a contrived one at that—and then attempting to explain by means of it the most diverse and unlimited phenomena.

In my own theory of color I have used the analytic method, arranging all possible known phenomena in a particular sequence, in order to see to what extent something universal might exist here, and under what rubric the phenomena might be classified; and I believe that I have in this way done some groundwork for the synthetic work which is the special obligation of the nineteenth century.

I have followed a like procedure in explaining all phenomena resulting from double reflection, and can transmit both pieces of work to an immediate or remote future with the knowledge of having returned these investigations to Nature and of having restored their rightful freedom.

We turn to another more general observation: a century that restricts

---

[31] First printed in the posthumous works of Goethe, 1833. See "Bibliographical Note," p. 255.
[32] French metaphysician and philosopher, 1792–1867. We find several complimentary references to Cousin in Eckermann's *Conversations of Goethe* for 1829. From this it is concluded that the essay was written in that year.

itself to analysis and seems almost afraid of synthesis, is on the wrong path; for only the two together, as with inhalation and exhalation, comprise the life of science.

A false hypothesis is better than none at all. The fact that it is false does not matter so much. However, if it takes root, if it is generally assumed, if it becomes a kind of credo admitting no doubt or scrutiny—this is the real evil, one which has endured through the centuries.

The Newtonian theory might be cited as an example. In its own time its deficiencies had already been held up to view, but Newton's outstanding attainments in general and his prestige in lay and professional circles, prevented opposition. It is the French especially who have the greatest blame in the spread and ossification of his theory, and in order to right their error, they should sponsor in the nineteenth century a fresh analysis of that complicated and ossified hypothesis.

The chief thing lost sight of in an exclusive application of analysis, is that all analysis presupposes synthesis. A sand heap cannot be analyzed. However, if it consists of various parts, let us say sand and gold, the washing of it would be an analysis in which the light substance is flushed away and the heavy substance held back.

Thus modern chemistry is engaged chiefly in separating what Nature has joined together. We annul Nature's synthesis in order to become aquainted with her in separated elements.

What higher synthesis could exist than a living entity? Yet how we have to torture ourselves with anatomy, physiology, and psychology to acquire only a limited concept of the whole complex, one that is always instantaneously restored, no matter how many parts it has been torn into.

A great danger encountered by the analyst is the application of his method in places where no synthesis exists. In such a case his efforts are truly like those of the Danaides, and we are witness to the saddest examples of this. For at bottom the analyst is really doing his work in order to arrive again at a synthesis. If none exists at the basis of the subject he is treating, he is working in vain to discover it. Observations are then only a hindrance to him, the more they increase in number.

The analyst must therefore take into primary consideration, or rather have ever in mind, the question whether he is really working with a mysterious synthesis or whether the subject he is treating is only an aggregation, a contiguity, a jointness, or whatever other modification of the idea we can think of. Those chapters of science in which no headway is achieved are subject to a suspicion of this sort. On this subject one might make very fruitful observations with respect to geology and meteorology.

Excursus[33]

THE FOLLOWING ESSAYS have as little right as the preceding ones to be regarded as finished literary work. Conceived individually from alternating points of view, under the influence of contradictory states of mind, written down at different times, they could not possibly form an integrated whole. The dates are not always given, partly because they had not always been noted down, and partly because I was able, as editor of my own papers, to eliminate such superfluous things as dates, and many other disagreeable things as well! Nevertheless, some things remain for which I do not answer: namely, the contradictions and repetitions I could not eliminate without wholly destroying what was inseparably linked with them.

And yet these notes, as parts of a human life, may serve as evidence of the infinitely greater number of phases an individual must pass through if, instead of continuing on a more ordinary life course, he feels impelled to develop himself universally, choosing as his motto:

> If you desire to reach out into the infinite,
> Move in all directions in the finite.[34]

Or as someone else has expressed it:

> Natura infinita est
> Sed qui symbola animadvertit,
> Omnia intelliget
> Licet non omnino.[35]

Friendly Gesture[36]

IN CLOSING, I cannot refrain from mentioning a satisfaction I have experienced repeatedly of late: I feel at one with scientists at home and abroad, in harmonious agreement with active, serious scholars who confess and maintain that one should grant the ultimate mystery of life but set no other bounds for scientists.

Must I not also grant and assume my own individuality, without ever

---

[33] This excursus or digression forms the introduction to *Natural Science in General; Morphology in Particular*, Vol. I, No. 2 (1820).

[34] Original with Goethe.

[35] A passage from Thomas Campanella: Nature is infinite, but those who give heed to her symbols will understand everything, even though not entirely.

[36] Written in 1820; printed in *Natural Science in General; Morphology in Particular*, Vol. I, No. 3 (1820).

actually comprehending my own nature; do I not constantly study myself without ever understanding either myself or others? And yet one goes merrily on and on.

Thus too with the world! It may stretch before us, without beginning, without end; the distance boundless, the near-at-hand impenetrable. Be it so—yet the depth and the extent to which the human intellect is capable of probing into its own mysteries and those of the world can never be defined or terminated.

Plea for Unity and Co-operation[37]

I CANNOT REFRAIN from mentioning the gratitude I felt on reading the following passage in Supplement 47 of the *Jena Literary Gazette*, 1821:

"Nees von Esenbeck's *Handbook of Botany*[38] joins the botanical efforts of Goethe, Steffens,[39] Schelver,[40] Oken,[41] Kieser,[42] and Wilbrand:[43] for these men give evidence, each in his own way, of the same inspiration. But who would care to make anxious inquiry into the specific debt we owe to one or to the other, or even to selfishly establish his own priority, treating the knowledge gained as though it were some materialistic and dead possession? Instead, each should give thanks to the divine guidance which has directed so many into the same school of thought and has bestowed upon our times the invaluable gift of harmonious cooperation of various intellects!"

This appeal, advising and urging unanimity in the treatment of the genuine and the true, is the beginning of the fulfillment of wishes I expressed under the title, "Meteors of the Literary Sky."[44] May they be brought closer, the Good Genius willing!

We know that fellow believers in a religious faith must renounce their personal idiosyncrasies (though not their individualities). So too in higher

---

[37] Written in 1821–1822; printed in *Natural Science in General; Morphology in Particular*, Vol. I, No. 4 (1822).

[38] Appeared in two parts, Nürnberg, 1820–1821. The author sent Goethe a copy of the first volume in November, 1820. For other references to von Esenbeck, see pp. 104, 200, 201.

[39] Heinrich Steffens, 1773–1845, poet, natural philosopher, physicist.

[40] See pp. 105, 106, 119, 189, 199.

[41] See p. 201. Oken first presented his system in *Textbook for a System of Natural Philosophy*, 1810.

[42] See pp. 199–200.

[43] See pp. 120–121.

[44] Published in *Natural Science in General; Morphology in Particular*, Vol. I, No. 2 (1820).

science, in the divine domain of Nature: we can co-operate effectively, we can recognize the constitution of the domain, we may participate in its control, but we can do this only if we, its citizens, patriotically renounce our idiosyncrasies and submerge ourselves in the whole, so that our individual contributions will completely disappear within the whole, hovering over posterity, in company with thousands of others, transfigured.

NATURE[45] (A FRAGMENT)

NATURE! We are encompassed and embraced by her—powerless to withdraw, yet powerless to enter more deeply into her being. Uninvited and unforewarned, we are drawn into the cycle of her dance and are swept along until, exhausted, we drop from her arms.

She is creating new forms eternally. What is now, has never been; and what has been, will never be again. All is new, yet ever the same.

We live within her, yet are foreign to her. Conversing with us endlessly, she never divulges her secret. We influence her continually, yet have no power over her.

She seems to stake everything on individuality, yet sets small value on the individual. She is ever building, ever tearing down, and her workshop is inaccessible.

She lives only in her children, yet where can they find her—their mother?

She is the supreme artist: with the simplest materials she creates the most remarkable contrasts; seemingly without effort she achieves perfection, yet her utmost precision is hidden by softness.

Each of her creations has its own being, each represents a special concept, yet together they are one.

She is putting on a spectacle, but whether she is watching it we cannot tell. But she is producing it for us who stand in the wings.

She is eternal life, eternal becoming, eternal change, yet she does not

---

[45] This composition appeared first in the winter of 1782–1783, in the *Tiefurter Journal*, a magazine in manuscript form which circulated among the associates of the Weimar Court, all contributions to which were anonymous. In March, 1783, Goethe denied its authorship in a letter to a friend. Years later, in 1828, when a copy of the journal was found among the Duchess Amalia's papers, he had quite understandably forgotten his earlier denial, and now stated that the content was in accord with his views at the time of its publication. The essay, usually included in editions of Goethe's scientific works, is of value in tracing the development of Goethe's ideas, and achieves point through the comment he wrote on it in a letter to a friend, given on pp. 244–245.

move forward. She ever transforms herself, without pausing to rest. She is constant, yet impatient with anything static, and has set her curse on stagnation. Her pace is measured, her exceptions few, her laws immutable.

She has pondered deeply and meditates incessantly—not as a human being but as Nature. By merely watching her we cannot fathom the mysterious final truth she is withholding.

Mankind exists in her and she in all mankind. She plays a friendly game with him, rejoicing all the more when he triumphs. Sometimes she carries on the game so enigmatically, before he is aware of it the game is ended.

Nature is even the unnatural. Those who cannot see her everywhere will not see her clearly anywhere. Even the crudest mediocrity is tinged with her genius.

She is enamored of herself, adoring herself with countless eyes and hearts. For self-enjoyment she dissects herself. Never weary of flaunting herself, she creates new beings to admire her.

She delights in illusion. As the harshest tyrant she will punish whoever destroys it in himself or in others, but whoever follows her trustingly she will press to her heart as her child.

The number of her children is infinite. With none is she miserly, but she has favorites to whom she is generous, for whom she will sacrifice much. She bestows her protection on greatness.

She pours her creations forth from the void, telling them neither whence they have come nor whither they are bound. Each must simply run his course, she alone knows the way.

Her mainsprings are few but they are never run down; they are ever efficient, ever diverse.

Since she is always creating new spectators, her spectacle is forever new. Life is her finest invention, and her device for producing an abundance of life is her masterstroke—death.

She shrouds man in darkness, yet drives him eternally on toward the light. She keeps him earthbound, lazy, and leaden, yet continually prods him awake from his torpor.

Because she loves forward movement, she endows man with longing, miraculously gaining much with so little! Longing becomes a blessing, quickly satisfied, quickly awakened again. Further longing becomes but a new source of pleasure. But balance is soon restored.

She continually sets out upon some distant goal and is continually arriving at her destination.

She is vanity itself, yet to us is of utmost importance.

She permits any child to experiment with her, any fool to judge her, and allows thousands to pass her by, apathetic and unseeing. Yet she draws pleasure and profit from all.

By resisting her laws we obey them. When we are intent upon working against her we are most in harmony with her.

All her gifts become benefits, since she has already made them necessities. She tarries that we may know longing; she hastens that we may not become surfeited.

She has neither language nor voice, but creates tongues and hearts through which she may speak.

Love is her crown. Through love may we know her. She has separated her creations by cleavages, yet in them is the urge to draw close. She has put them apart that she may draw them together. As reward for a lifetime of labor she grants them a few draughts from the goblet of love.

She is the All-in-One. She metes out her own reward and punishment. She rejoices in herself and tortures herself. She is at once harsh and gentle, revolting and beguiling, helpless and all-powerful. Within her all things exist forever. She knows neither past nor future. The present is her eternity.

She is gracious and I praise her work. She is wise and serene. We wrest no secrets from her, extort no gifts, receiving only those which she yields of her own free will. She is deceitful, but to a good end, and it is wise to ignore her cunning.

She is complete, yet ever unfinished. Her way of working can continue forever.

To each of us she appears in a different form. Disguised by thousands of names and definitions, she is still the same.

She has brought me here and will escort me hence. I entrust myself to her care; she may do with me as she pleases, for I know she will not despise me—her handiwork. Even now it is not I who am speaking of her. No. These are her words—both the true and the false. And hers is the blame and the glory for all.

(English rendering by Aldyth Morris)

COMMENTARY ON "NATURE"[46]

THE PIECE OF WRITING in question was given to me from among the papers of our late beloved Duchess Amalia. It is in handwriting well

---

[46] Goethe's letter to Chancellor F. T. A. H. von Müller, May 24, 1828.

known to me, that of a person upon whose services I often drew in the eighties.

I cannot recall actually writing these remarks, but they do agree with the ideas preoccupying my mind at that time. I should like to call my state of knowledge in that period a "comparative," compelled to orient itself toward a not yet attained "superlative." One sees in them an inclination toward a kind of pantheism, inasmuch as an inexplorable, undefined, humorous, self-contradictory entity is visualized at the base— a playful jester, one to be taken nevertheless in bitter earnestness.

The composition lacks the consummating concept of two of Nature's activating forces: polarity and progression. Polarity is a property of matter insofar as we conceive of it as material; progression is a property of spirit, insofar as we conceive of it as spiritual. The first is in continual attraction and repulsion, the latter in constant upward striving. But since matter never exists without spirit, and spirit never without matter, matter is capable of advancing and spirit has the power to attract and repulse. We have an analogy in the fact that only an individual who has analyzed sufficiently is in a position to do the thinking prerequisite to synthesis, and only one who has sufficiently synthesized, is in a position to make a reanalysis.

At the period when I might possibly have written the composition, I was chiefly occupied with comparative anatomy, taking enormous pains in 1786 to interest others in my conviction that the intermaxillary bone was undeniably present in man. Even men of intelligence were unwilling to admit the importance of this statement; its truth was denied by the best observers, and I was compelled, as in so many other instances, to quietly continue my way alone.

I ceaselessly pursued the versatility of Nature in the plant world and in 1788, in Sicily, by means of active observations and creative imagination, I was successful in evolving the theory of metamorphosis of plants. The metamorphosis of animals was closely related, and in 1790, in Venice, the origin of the skull from the vertebrae was revealed to me. I now pursued the construction of the animal type more zealously, dictated the outline of my theory to Max Jacobi in Jena in 1795, and soon had the joy of seeing myself succeeded in this field by other German scientists.

If one recalls the splendid development of this idea, through which all natural phenomena have gradually been linked together for the human intellect, and if one then carefully rereads the essay here referred to, one can smilingly contrast the comparative, as I have called it, with the superlative achieved, and rejoice in fifty years of progress.

Appendix

Biographical Notes

THESE NOTES mention only the chief influences and events in an extraordinarily rich and varied life, and only the most important literary works from more than 130 large volumes comprising the Weimar edition of Goethe's works.

1749   Johann Wolfgang Goethe born in Frankfurt. Son of Johann Kaspar Goethe, Dr. jur., Imperial Councillor; and Katharina Elizabeth, nee Textor. Sister: Cornelia.

1765–1768   Student days in Leipzig.
         Nominally a student in law, he devotes his time to art, science, and belles-lettres. Writes his first important literary works: a volume of poems, *Annette,* and a pastoral play, *Die Laune des Verliebten* (The Moody Lover).
         Falls seriously ill.

1768   Recovers from illness in parental home.
         Fräulein von Klettenberg, a Moravian pietist, friend of Goethe's mother, has a profound spiritual influence upon him, as mirrored later in "The Confessions of a Beautiful Soul," a section of the novel *Wilhelm Meisters Lehrjahre* (Wilhelm Meister's Apprenticeship).

1770–1771   As student in Strassburg, devotes more time to lectures on science and medicine than to his chosen field of law. Undergoes powerful intellectual awakening under the influence of theologian and philosopher Johann Gottfried Herder (1744–1803), who translated folk songs from many languages, was among the first to mark out the lines of evolution, and was one of the first Shakespeare enthusiasts in Germany. (Later, through Goethe's influence, Herder became court preacher at Weimar and spent the remainder of his life there.)
         Falls in love with Frederica Brion, who becomes the inspiration for some of his most beautiful lyrics, and the model for the figures of Marie in *Götz von Berlichingen* and of Gretchen in *Faust.*

1771   Receives his licentiate in law and settles down to practice in his native city of Frankfurt.

1772   Practices law in Wetzlar, in the Imperial Chancery.
         His love for Charlotte Buff inspires the character of Lotte in *Die Leiden des jungen Werther* (The Sorrows of Young Werther).

1773	The appearance in print of *Götz von Berlichingen*, establishes Goethe as the chief representative of Storm and Stress, a literary movement which grew up as a reaction against the mannered literature of the Enlightenment.
1774	On a journey to the Rhine, Goethe makes the acquaintance of the Hereditary Prince of Weimar, Karl August. Goethe works on his *Urfaust*, nucleus of the later *Faust*. Publishes the sentimental novel, *Die Leiden des jungen Werther*, which takes Germany and the world by storm. (Napoleon carried a copy of it on his campaigns, and it became the first German work to be translated into the Chinese.)
1775	Becomes secretly engaged to Lili Schönemann; engagement broken at her request the same year. Journeys to the Rhineland with Lavater, the physiognomist, who first stimulates Goethe to serious scientific research, and to whose extensive work on physiognomy Goethe contributes the chapter on animal skulls. Accepts the invitation to visit Karl August, who had now assumed an active role in the governing of Weimar. Goethe arrives on November 7, intending to pay only a short visit, but is destined to spend the rest of his life there, and to form connections of decisive influence on his life.
1775–1785	A momentous decade in the life of Goethe. The attachment and admiration of the young Duke for Goethe is so great that he wishes him to be admitted to public service. In spite of strong opposition from some high officials, Goethe is appointed Privy Councillor in the spring of 1776. Goethe attends meetings regularly and performs many special services for the Duke as Minister of Finance, Minister of War, Minister of Mines, Minister of Arts. In the latter capacity he concerns himself with public education in Weimar and with the affairs of the university in Weimar's sister city of Jena. During this first decade in Weimar, Goethe devotes much time to natural science, chiefly osteology and botany. The scientific collections of the university in Jena are open to him, and he has unlimited opportunity for consulting university scientists. In the course of his osteological studies he discovers the intermaxillary in humans, in 1784, thereby paving the way for the theory of evolution. In botany, he is able to carry on research in connection with his administration of the ducal gardens and of the forests of the duchy. In connection with his work with the mines in Ilmenau, he pursues geological studies. Goethe's extensive work as statesman and scientist inevitably restricts his literary output. None of his larger projected works is completed. However, his devotion to Charlotte von Stein throughout this period inspires some of the loveliest lyrics he ever wrote. Her influence also finds its reflection in Goethe's

BIOGRAPHICAL NOTES

drama *Iphigenie* embodying the classic ideals of integrity and self-control.

1782 — Goethe is elevated to the nobility.

1785 — Goethe spends the summer in Karlsbad for the cure, botanizes extensively.

1786–1788 — The Italian journey.

Goethe's decision to leave Weimar comes as the culmination of a long-cherished wish to visit Italy, and from the desire to be relieved of the burdens of state. First obtaining the consent of his duke, he leaves without informing his friends, not even Charlotte von Stein. In Italy Goethe sojourns in Rome, where he studies ancient art under the tutelage of a circle of German artists, and is won over to classic ideals of art and literature. The lush vegetation of Italy spurs his interest in botany, and in Naples he is inspired with the idea embodied in his essay on the metamorphosis of plants, namely, that all plant organs are but modifications of the embryo leaves, the cotyledons.

1787–1790 — Goethe's collected works are published by Göschen, in eight volumes.

1788 — Goethe returns to Weimar, greatly enriched and stimulated by his Italian sojourn, but is nevertheless destined to experience great unhappiness. His old acquaintances, piqued by his unannounced departure for Italy, are now further estranged by his enthusiasm for events and subjects unfamiliar to them. Frau von Stein's coldness, and the subsequent break between them, form another source of unhappiness. In addition, his interest in science is offensive to many of his friends, who feel that this interest is interfering with his poetic activity, the scientific pursuits becoming interlinked in their minds with the estrangement.

The alienation becomes still greater when Goethe enters into a "conscience marriage" with Christiane Vulpius, a woman of lower social standing than his own. Her high spirits, gaiety, charm, and warmth are a source of consolation to him during this period, the rapport between them being reflected in *Die Römischen Elegien* (Roman Elegies).

At his request, Goethe is relieved of most of his administrative tasks, but retains the supervision of the Institute of Arts and Sciences, duties which he has found congenial and helpful in attaining his literary and scientific goals.

1789 — Goethe's son August is born.

1790 — On his second Italian journey he evolves the vertebral theory of the skull.

After his return from Italy, Goethe travels to Silesia at the request of Karl August, now a major general with the Prussian army. Keeping to himself in camp, Goethe devotes much time to botany and mineralogy, and begins an essay on the development of animals.

His essay on the metamorphosis of plants is published; also his drama *Tasso*.

1791	The Weimar Court Theatre is opened under Goethe's direction.
1791–1792	*Beiträge zur Optik* (Contributions to Optics) is published.
1792	Goethe accompanies Karl August on a military campaign to France.
1793	Goethe joins Karl August during the allies' siege of Mayence.
1794–1805	Period of friendship with Schiller, "one of the noblest relationships granted me in my later years."

Although Schiller had come to Weimar for the express purpose of making Goethe's acquaintance, Goethe had felt repelled at their first meeting. Ten years older than Schiller, he had just emerged from the extravagances of the Storm and Stress, in which Schiller, currently feted as the author of *The Robbers*, was still involved. Efforts of mutual friends to effect a conciliation were without result. In July, 1794, such a conciliation takes place naturally as the two poets meet by chance at a scientific meeting and become involved in a discussion of Goethe's theory of plant metamorphosis.

There now begins their famous literary friendship. With almost daily correspondence between them, there is a fruitful interchange of opinions and influences: for instance, it was Schiller who induced Goethe to resume work on *Faust*; it was Goethe who spurred Schiller on to writing *Wilhelm Tell*.

1795	*Wilhelm Meisters Lehrjahre* (Wilhelm Meister's Apprenticeship) is published.
1797	Plan for *Faust* is written down. Completion of *Hermann und Dorothea*, an epic idyll against the background of the French Revolution, is completed.
1798–1805	The rebuilt Weimar Court Theatre opens in 1798, and there follows flourishing period with performances of plays by Goethe and Schiller.
1801	Goethe falls seriously ill.
1803	Death of Herder.
1805	Death of Schiller.
1806	Weimar is plundered by the French, after the Battle of Jena in October. Goethe is placed in danger, which Christiane averts. Touched by her spirited behavior in his defense and realizing her

need in such troubled times for the legal protection of his name, Goethe decides to legalize their marriage.

1808 *Faust* I is published.
Napoleon, in nearby Erfurt for the Congress of Princes, commands an audience with Goethe. It was in Erfurt that Napoleon made his famous remark about Goethe, "*Voilà un homme!*"

1809 His novel *Die Wahlverwandtschaften* (Elective Affinities) is published.

1810 *Farbenlehre* (Theory of Colors) is published.

1811 First part of Goethe's autobiography *Dichtung und Wahrheit* (Truth and Poetry) appears. Publication extends through 1831.

1812 Goethe meets Beethoven for the first time in Teplitz. Though each appreciated the other's greatness, and Beethoven had long admired Goethe from a distance and had been desirous of a meeting, the temperaments of the two men differed so greatly that closer relations never resulted.

1814 On his journey to the Rhine writes the poem cycle, *West-Oestlicher Divan* (West-Eastern Divan), inspired by his love for the poetess Marianne Willemer.

1816 Death of Christiane.

1817 Goethe gives up his directorship of the Weimar Theatre. Marriage of Goethe's son August to Ottilie von Pogwisch, who takes over the direction of the household.

1823 The *Marienbader Elegie* (The Marienbad Elegy), inspired by the aging poet's love for young Ulrike von Levetzow, and her refusal of his offer of marriage, gives expression to the realization that he is growing old and that little time remains for finishing his literary work.
Johann Peter Eckermann visits Weimar and enters Goethe's employ as secretary. His gift for engaging Goethe in conversation and then reproducing these conversations, with extraordinary fidelity, in his diaries is reflected in his *Gespräche mit Goethe* (Conversations with Goethe), published in 1837 and after.

1823–1832 During the last decade of Goethe's life, hundreds of distinguished visitors from Germany and abroad are drawn to Weimar to interview him.

1825 Goethe resumes work on *Faust* II, designating it his chief task in the period to follow.
Goethe is honored by Jubilee, celebrating the fiftieth anniversary of his arrival in Weimar.

1827	Death of Frau von Stein.
1828	Death of Karl August.
1829	Goethe's eightieth birthday is officially celebrated. *Faust* is performed for the first time in Weimar. *Wilhelm Meisters Wanderjahre* (Wilhelm Meister's Travels) is completed.
1830	Death of Goethe's son August in Rome.
1831	Goethe makes his will, designating Eckermann his literary executor. *Faust* II is completed in August; also the last installment of his autobiography. The essay on metamorphosis is republished with an interleaved French translation by the Swiss scientist Frederic Soret, 1795–1865, who had been called to Weimar in 1822 to act as tutor to Prince Karl Alexander. The translation was executed under the direct supervision of Goethe.
1832	Goethe's death on March 25.

Bibliographical Note

THE GOETHEAN EDITIONS cited in the footnotes are as follows:

J. W. von Goethe Herzoglich Sachsen-Weimarischen Geheimraths Versuch die Metamorphose der Pflanzen zu erklären (Gotha: bei Carl Wilhelm Ettinger, 1790). First edition of the essay on the metamorphosis of plants.

J. W. von Goethe. *Versuch über die Metamorphose der Pflanzen*, Uebersetzt von Friedrich Soret, nebst geschichtlichen Nachträgen (Stuttgart: in der Cotta'schen Buchhandlung, 1831), with parallel title: J. W. de Goethe: *Essai sur la Métamorphose des plantes*, traduit, par Frédéric Soret, et suivi de notes historiques (Stuttgart: J. G. Cotta, Libraire, 1831).

Goethes Werke, Vollständige Ausgabe letzter Hand (Stuttgart and Tübingen: in der J. G. Cotta'schen Buchhandlung, 1830–1833).

Goethes Werke, herausgegeben im Auftrage der Grossherzogin Sophie von Sachsen (Weimar, Hermann Böhlau). I. Abteilung, *Goethes Werke*, 55 vols., 1887–1918; II. Abteilung, *Goethes naturwissenschaftliche Schriften*, 13 vols., 1890–1904; III. Abteilung, *Goethes Tagebücher*, 13 vols., 1887–1903; IV. Abteilung, *Goethes Briefe*, 50 vols., 1887–1912. Commonly known as the "Weimar Edition."

Zur Naturwissenschaft überhaupt, besonders zur Morphologie. Erfahrung, Betrachtung, Folgerung, durch Lebenzereignisse verbunden [Natural Science in General; Morphology in Particular. Experiences, Observations, Conclusions, Together with Events from My Life.] (Stuttgart und Tübingen; in der J. G. Cotta'schen Buchhandlung). Band I: Nr. 1, 1817, Nr. 2, 1820; Nr. 3, 1820; Nr. 4, 1822. Band II: Nr. 1, 1823; Nr. 2, 1824.

TABLE OF CONTENTS[1]

VOLUME I, No. 1 (1817)

* Our Undertaking Is Defended
* Our Objective Is Stated
* The Content Is Given a Foreword
  The History of My Botanical Studies[2]
* Genesis of the Essay on the Metamorphosis of Plants
* The Metamorphosis of Plants

---

[1] The starred essays are presented in this translation, though not in the original order. (See Translator's Preface.)

[2] An extended version of this essay, entitled "The Author Relates the History of His Botanical Studies," is presented in this translation.

* History of the Manuscript
* History of the Brochure in Print
* Discovery of a Worthy Forerunner
* Propitious Encounter

VOLUME I, No. 2 (1820)

Orphean Prophecy
* Excursus
* Influence of the New Philosophy
* Intuitive Judgment
* Indecision and Surrender
* Creative Urge
* Three Favorable Reviews
* Other Friendly Overtures
* Later Studies and Collections
Outlines of Osteology
The Intermaxillary Bone
Addenda
In Memory of Caspar Friedrich Wolff

VOLUME I, No. 3 (1820)

Lectures on the Three First Chapters of a Projected Work on Comparative Osteology
* Pollination, Volatilization, and Exudation
* Friendly Gesture
Angry Gesture

VOLUME I, No. 4 (1822)

By Way of Introduction
* Botany[3]
* Remarkable Healing of a Badly Injured Tree
* Notes for an Essay on Plant Culture in the Duchy of Weimar
Zoology
* Analogous Volatilization
Sloths and Pachyderms
Dr. Carus: On Opinions Concerning Shell and Bone Structure
Fossil Steer
* *Color Chart of Organic Nature,* by Wilbrand and Ritgen (Book Review)
* *History and Development of the Plant World,* by Schelver (Book Review)
Luke Howard to Goethe
Remarks

---

[3] In this translation, the contents of the essay are presented in two sections: "Plea for Unity and Co-operation" and "Increasing Difficulty of Botanical Instruction."

Volume II, No. 1 (1823)

* An Analogous Procedure
  Remarks on a Collection of Diseased Ivory
  Carus: On Primitive Forms of *Pelecypoda* and *Gastropoda*
* Problems
* Considerable Assistance from One Ingeniously Chosen Word
  On Requirements for Illustrations for Natural History in General and for Osteology in Particular
* *Charts of Organic Nature,* by Wilbrand and Ritgen (Book Review)
* *Nature, Its System and History,* by Voigt (Book Review)

Volume II, No. 2 (1824)

Carus: Outlines of General Natural Science
On Hops and Their Disease Called Smut
* On Smut, Blight, and Honeydew[4]
* Additional Notes on Smut on Hops[4]
Illustrations of Outstanding Horses in the Royal Prussian Studs
Wrong Roads Taken by a Botanist
* *Genera et Species Palmarum,* by Martius (Book Review)

Texts from which the translations were made:

Goethes Werke. Herausgegeben von Prof. Dr. Karl Heinemann. Kritisch durchgesehene and erläuterte Ausgabe. 29. Band und 30. Bearbeitet von Wilhelm Bölsche (Leipzig und Wien: Bibliographisches Institut, 1908). This text was the one used throughout with the exception of the essay mentioned in the following item.

Goethes morphologische Schriften, ausgewählt und eingeleitet von Wilhelm Troll (Jena: Diederich, 1932). The essay entitled "Preliminary Notes for a Physiognomy of Plants" was translated from this text. The illustrations for the essay on the metamorphosis of plants were also taken from this text, with the permission of the publishers.

---

[4] The material of these two essays appears in this translation under the title "Pollination, Volatilization, and Exudation."

Selected Bibliography

ARTICLES

Arber, Agnes. "Goethe's Botany," *Chronica Botanica*, X, No. 2, 67–124.

Bartlett, Harley H. "Goethe as a Biologist," *Michigan Alumnus Quarterly Review*, LV, No. 24, 300–312.

Bloch, Robert. "Goethe, Idealistic Morphology, and Science," *American Scientist*, XL, No. 2 (April, 1952), 313–322.

Engard, Charles Joseph. "Poetic Scientist," *Science Monthly*, LXVIII (May, 1949), 305–309.

Lange, V. "Goethe: Science and Poetry," *Yale Review*, n.s. XXXVIII, No. 4 (June, 1949), 623–639.

Meyer, Heinrich. "Goethe as Scientist. A Problem in Historical Method," *Monatshefte für deutschen Unterricht*, XLI (December, 1949), 415–423.

BOOKS

Bielschowsky, Albert. *The Life of Goethe*, auth. tr. by William Cooper (New York and London: G. P. Putnam's Sons, 1905–1908), III, ch. 3, "The Naturalist," by Salomon Kalischer.

Lewes, George Henry. *The Life of Goethe* ("Everyman's Library"; London and New York: D. M. Dent & Sons, Ltd., 1908), esp. "The Poet as a Man of Science," pp. 336–377. Work first published in 1855.

Magnus, Rudolf. *Goethe as a Scientist*, tr. Heinz Norden, with foreword by Günther Schmid (New York: Henry Schuman, 1948). German title: *Goethe als Naturforscher* (Leipzig: 1906).

Sherrington, Sir Charles Scott. *Goethe on Nature and Science* (2nd ed.; Cambridge, England: Cambridge University Press, 1949).

Steiner, Rudolf. *Goethe's Conception of the World* (New York: Anthroposophic Press, 1928).

Vietor, Karl. *Goethe the Thinker*, tr. Bayard Q. Morgan (Cambridge: Harvard University Press, 1950).

DATE DUE

Demco, Inc. 38-293